T0302181

MXene-Based
Photocatalysts

Emerging Materials and Technologies
Series Editor
Boris I. Kharissov

Nanomaterials for Water Treatment and Remediation
Srabanti Ghosh, Aziz Habibi-Yangjeh, Swati Sharma, and Ashok Kumar Nadda

2D Materials for Surface Plasmon Resonance-Based Sensors
Sanjeev Kumar Raghuwanshi, Santosh Kumar, and Yadvendra Singh

Functional Nanomaterials for Regenerative Tissue Medicines
Mariappan Rajan

Uncertainty Quantification of Stochastic Defects in Materials
Liu Chu

Recycling of Plastics, Metals, and Their Composites
R.A. Ilyas, S.M. Sapuan, and Emin Bayraktar

Viral and Antiviral Nanomaterials: Synthesis, Properties, Characterization, and Application
Devarajan Thangadurai, Saher Islam, and Charles Oluwaseun Adetunji

Drug Delivery Using Nanomaterials
Yasser Shahzad, Syed A.A. Rizvi, Abid Mehmood Yousaf, and Talib Hussain

Nanomaterials for Environmental Applications
Mohamed Abou El-Fetouh Barakat and Rajeev Kumar

Nanotechnology for Smart Concrete
Ghasan Fahim Huseien, Nur Hafizah A. Khalid, and Jahangir Mirza

Nanomaterials in the Battle Against Pathogens and Disease Vectors
Kaushik Pal and Tean Zaheer

MXene-Based Photocatalysts: Fabrication and Applications
Zuzeng Qin, Tongming Su, and Hongbing Ji

Advanced Electrochemical Materials in Energy Conversion and Storage
Junbo Hou

For more information about this series, please visit: https://www.routledge.com/
Emerging-Materials-and-Technologies/book-series/CRCEMT

MXene-Based Photocatalysts

Fabrication and Applications

Edited by
Zuzeng Qin, Tongming Su, and Hongbing Ji

CRC Press is an imprint of the
Taylor & Francis Group, an **informa** business

First edition published 2022
by CRC Press
6000 Broken Sound Parkway NW, Suite 300, Boca Raton, FL 33487-2742

and by CRC Press
4 Park Square, Milton Park, Abingdon, Oxon, OX14 4RN

© 2022 Taylor & Francis Group, LLC

CRC Press is an imprint of Taylor & Francis Group, LLC

Library of Congress Cataloging-in-Publication Data
Names: Qin, Zuzeng, editor. | Su, Tongming, editor. | Ji, Hongbing, editor.
Title: MXene-based photocatalysts : fabrication and applications / edited by Zuzeng Qin,
Tongming Su, and Hongbing Ji.
Description: First edition. | Boca Raton : CRC Press, 2022. | Series: Emerging materials and
technologies | Includes bibliographical references and index. | Summary: "MXene-Based
Photocatalysts: Fabrication and Application introduces readers to the fundamentals,
preparation, microstructure characterization, and a variety of applications of MXene-based
photocatalysts. This work serves as an invaluable guide to advanced students, industry
professionals, professors, and researchers in materials science and engineering, studying
photocatalysis, energy, and environmental applications"-- Provided by publisher.
Identifiers: LCCN 2021047662 (print) | LCCN 2021047663 (ebook) | ISBN
9780367742904 (hbk) | ISBN 9780367742911 (pbk) | ISBN 9781003156963 (ebk)
Subjects: LCSH: Photocatalysis--Materials. | MXenes. | Two-dimensional materials.
Classification: LCC QD716.P45 M93 2022 (print) | LCC QD716.P45 (ebook) | DDC 541/
.395--dc23/eng/20211130
LC record available at https://lccn.loc.gov/2021047662
LC ebook record available at https://lccn.loc.gov/2021047663

ISBN: 978-0-367-74290-4 (hbk)
ISBN: 978-0-367-74291-1 (pbk)
ISBN: 978-1-003-15696-3 (ebk)

DOI: 10.1201/9781003156963

Typeset in Times
by MPS Limited, Dehradun

Contents

Preface

Photocatalysis with MXene-based photocatalysts have attracted great attention in the field of transforming solar energy into chemical energy, which has demonstrated a huge potential in alleviating the problems of energy crisis and environmental pollution. MXene, a two-dimensional transition metal carbide, nitride, and carbonitride discovered in 2011, exhibits great potential in various applications, such as photocatalysis, sensing and biosensing, electrochemical energy storage, energy conversion and storage, catalysis, rechargeable batteries, supercapacitors, biomedicine, and environmental pollution. Among them, MXene can be used as co-catalyst of semiconductors for photocatalysis due to having a hydrophilic surface, large specific surface area, and high electronic conductivity. Moreover, the suitable Fermi level of MXene is beneficial to form the Schottky junction with the semiconductor, which can enhance the separation of photogenerated electrons and holes. Consequently, MXene-based photocatalysts shows a great prospect in the field of photocatalysis.

In this book, the preparation method, morphology, and property of MXene, the construction of MXene-based photocatalysts, and the application of MXene-based photocatalysts in photocatalytic water-splitting, photocatalytic CO_2 reduction, photocatalytic degradation, and other fields are summarized. In addition, the ongoing challenge and opportunity for the future development of MXene-based photocatalysts in this exciting yet still emerging area of research was proposed. We believe that this instructive book can serve as an invaluable guide to the junior and senior undergraduate students, graduate students, industry professionals, professors, scientists, and researchers working in the field of photocatalysis, materials, energy, and environmental science.

Zuzeng Qin
Tongming Su
Hongbing Ji

Editors

Zuzeng Qin, PhD, earned a bachelor's and PhD in chemical engineering and technology at Guangxi University in 2000 and 2009, respectively, and worked as a postdoc at Sun Yat-Sen University from August 2012 to March 2015, and went to the University of Tennessee, Knoxville as a visiting scholar from October 2015 to October 2016. His research interests include environmentally friendly catalytic processes, the synthesis/separation of fine chemicals, and 2D materials for environmental and energy. He is the Vice-Dean of the School of Chemistry and Chemical Engineering at Guangxi University and an editorial board member of *Advanced Composites and Hybrid Materials*.

Tongming Su, PhD, earned a PhD in industrial catalysis at the School of Chemistry and Chemical Engineering, Guangxi University, China in 2018. From September 2016 to March 2018, he conducted scientific research on MXene-based materials at the Oak Ridge National Lab (ORNL) and the University of Tennessee, Knoxville as a joint PhD student supported by the China Scholarship Council. He is an Assistant Professor at the School of Chemistry and Chemical Engineering at Guangxi University, China. His research interests focus on the synthesis of MXene-based materials and their application in catalysis, energy conversion, and environmental remediation.

Hongbing Ji is a full professor from the School of Chemistry at Sun Yat-Sen University. He is the winner of the National Science Fund for Distinguished Young Scholars, the young and middle-aged leader of science and technology innovation of the Ministry of Science and Technology, the Distinguished Professor of the Pearl River Scholars of Guangdong Province, the Dean of the Fine Chemical Research Institute of Sun Yat-Sen University, the Dean of the Huizhou Research Institute of Sun Yat-Sen University, and an editorial board member of *Chinese Journal of Chemical Engineering*. His research interests are the green chemistry, green chemical technology and management of the chemical industry park.

Contributors

Liuyun Chen
School of Chemistry and Chemical
  Engineering
Guangxi University
Nanning, China

Ying-Hui Chin
Faculty of Petrochemical Engineering
  and Green Technology
Universiti Tunku Abdul Rahman
Perak, Malaysia

Man-Kit Choong
Faculty of Petrochemical Engineering
  and Green Technology
Universiti Tunku Abdul Rahman
Perak, Malaysia

Nongfeng Huang
School of Chemistry and Chemical
  Engineering
Guangxi University
Nanning, China

Hongbing Ji
School of Chemistry and Chemical
  Engineering
Guangxi University
Nanning, China
and
School of Chemistry
Sun Yat-Sen University
Guangzhou, China

Sze-Mun Lam
Faculty of Petrochemical Engineering
  and Green Technology
Universiti Tunku Abdul Rahman
Perak, Malaysia

Qingqing Li
School of Chemistry and Chemical
  Engineering
Guangxi University
Nanning, China

Xuan Luo
School of Chemistry and Chemical
  Engineering
Guangxi University
Nanning, China

Chengzheng Men
School of Chemistry and Chemical
  Engineering
Guangxi University
Nanning, China

Zuzeng Qin
School of Chemistry and Chemical
  Engineering
Guangxi University
Nanning, China

Jin-Chung Sin
Faculty of Petrochemical Engineering
  and Green Technology
Universiti Tunku Abdul Rahman
Perak, Malaysia

Tongming Su
School of Chemistry and Chemical
  Engineering
Guangxi University
Nanning, China

Jin-Han Tan
Faculty of Petrochemical Engineering
  and Green Technology
Universiti Tunku Abdul Rahman
Perak, Malaysia

Ya Xiao
School of Chemistry and Chemical
    Engineering
Guangxi University
Nanning, China

Zi-Jun Yong
Faculty of Petrochemical Engineering
    and Green Technology
Universiti Tunku Abdul Rahman
Perak, Malaysia

1 Introduction

Tongming Su and Zuzeng Qin
School of Chemistry and Chemical Engineering, Guangxi
University

Hongbing Ji
Fine Chemical Industry Research Institute, School of
Chemistry, Sun Yat-Sen University

CONTENTS

Heterogeneous photocatalysis is an innovative catalysis technology for trans-forming solar energy into chemical energy, which has demonstrated huge potential in alleviating the thorny problems of energy crisis and environmental pollution (Parrino et al. 2018; Su, Shao, et al. 2018; Su, Qin, et al. 2019; Li et al. 2021). In general, the photocatalytic process includes the following steps. First, under light irradiation, the photogenerated electron-hole pairs is produced on the photocatalysts when the energy of the incident photon is equal to or higher than the band gap of the semiconductor photocatalyst. Second, the photogenerated charge carriers are se-parated and transfered to the surface of photocatalyst. Finally, the surface-redox reactions occur on the surface-active sites of the photocatalyst by consuming the separated photoinduced charge carriers (Su, Shao, et al. 2018).

However, due to the fast recombination of the photogenerated electron-hole pairs and the inadequate surface-active sites of the photocatalyst, the photocatalytic per-formance of the semiconductor photocatalyst is seriously inhibited, which is still far from the requirements of practical application (Su et al. 2016). Therefore, great effort has been put into making the new-type photocatalysts that can effectively promote the separation of the photogenerated electrons and holes and provide more active sites, such as the Z-scheme system (Zhang, Mohamed, and Ong 2020), Type II hetero-structure (Jin et al. 2020), surface modification (Low, Cheng, and Yu 2017), metal and nonmetal doping (Wu et al. 2020), interface modification (Yang et al. 2021), crystal engineering (Bai et al. 2019), et al. Among them, the heterojunction photo-catalyst has attracted great attention in recent years because the heterojunction pho-tocatalyst can accelerate the transfer and the separation of the photoinduced electrons and holes, enhance the light-absorption capacity, and provide more active sites for photocatalytic reaction.

DOI: 10.1201/9781003156963-1

Among the heterojunction photocatalyst, MXene-based photocatalysts have obtained wide attention in the field of photocatalysis (Kuang et al. 2020). MXene, a two-dimensional transition metal carbide, nitride, and carbonitride, was discovered in 2011 (Naguib et al. 2011). In recent years, MXene, having the properties of high surface area, superb hydrophilic surface, great electrical conductivity, and excellent oxidation resistance, has been widely used as a cocatalyst of photocatalyst. Since the revolutionary work by researchers at Drexel University on the successful preparation of Ti_3C_2 MXene from exfoliation of Ti_3AlC_2 through wet-chemical etching in HF acid, various types of MXenes have been developed via several methods and used for varied photocatalytic application.

Due to the electrical conductivity, high surface area, and high work function of MXenes, MXenes have shown great potential as cocatalysts of semiconductor photocatalysts (Wang et al. 2020). When MXene is used as the cocatalyst, the photoinduced electrons can transfer from the semiconductor to the MXene materials; thus, the separation of the photogenerated electrons and holes can be significantly enhanced (Su, Hood, et al. 2019a). In addition, due to the hydrophilicity of the MXene, the MXene can easily combine with the semiconductor to form the MXene-based photocatalyst, and a compact interface can be formed between the MXene and the semiconductor; this further improves the separation of the photogenerated electron-hole pairs (Su, Peng, et al. 2018). Moreover, a Schottky junction can be formed at the MXene/semiconductor interface, which inhibits the backflow of the electron from the MXene to the semiconductor, thus suppressing the recombination of the photogenerated electrons and holes (Su, Hood, et al. 2019b). Finally, the two-dimensional structure and the high specific surface area of the MXene can expose more surface-active sites for the photocatalytic reaction (Chen et al. 2021). According to the above advantages, the MXene-based heterostructure is considered a promising photocatalyst for the practical application of photocatalysis.

This book summarizes the preparation method, morphology, and property of MXene, the construction of MXene-based photocatalysts, and the application of MXene-based photocatalysts in the photocatalytic water splitting, photocatalytic CO_2 reduction, photocatalytic degradation, and other fields. In addition, the ongoing challenge and opportunity for the future development of MXene-based photocatalysts in this exciting yet still emerging area of research will be proposed.

REFERENCES

Bai, Song, Chao Gao, Jingxiang Low, and Yujie Xiong. 2019. Crystal Phase Engineering on Photocatalytic Materials for Energy and Environmental Applications. *Nano Research* 12 (9):2031–2054.
Chen, Liuyun, Kelin Huang, Qingruo Xie, et al. 2021. The Enhancement of Photocatalytic CO_2 Reduction by the in Situ Growth of TiO_2 on Ti_3C_2 MXene. *Catalysis Science & Technology* 11 (4):1602–1614.
Jin, Chun, Wei Li, Yasi Chen, et al. 2020. Efficient Photocatalytic Degradation and Adsorption of Tetracycline over Type-II Heterojunctions Consisting of ZnO Nanorods and K-Doped Exfoliated g-C_3N_4 Nanosheets. *Industrial & Engineering Chemistry Research* 59 (7):2860–2873.

Kuang, Panyong, Jingxiang Low, Bei Cheng, Jiaguo Yu, and Jiajie Fan. 2020. MXene-Based Photocatalysts. *Journal of Materials Science & Technology* 56:18–44.

Li, Sumei, Saisai Shan, Sha Chen, et al. 2021. Photocatalytic Degradation of Hazardous Organic Pollutants in Water by Fe-MOFs and Their Composites: A Review. *Journal of Environmental Chemical Engineering* 9 (5):105967.

Low, Jingxiang, Bei Cheng, and Jiaguo Yu. 2017. Surface Modification and Enhanced Photocatalytic CO_2 Reduction Performance of TiO_2: A Review. *Applied Surface Science* 392:658–686.

Naguib, Michael, Murat Kurtoglu, Volker Presser, et al. 2011. Two-Dimensional Nanocrystals Produced by Exfoliation of Ti_3AlC_2. *Advanced Materials* 23 (37):4248–4253.

Parrino, F., M. Bellardita, E. I. García-López, G. Marcì, V. Loddo, and L. Palmisano. 2018. Heterogeneous Photocatalysis for Selective Formation of High-Value-Added Molecules: Some Chemical and Engineering Aspects. *ACS Catalysis* 8 (12):11191–11225.

Su, Tongming, Zachary D. Hood, Michael Naguib, et al. 2019a. 2D/2D Heterojunction of Ti_3C_2/g-C_3N_4 Nanosheets for Enhanced Photocatalytic Hydrogen Evolution. *Nanoscale* 11 (17):8138–8149.

Su, Tongming, Zachary D. Hood, Michael Naguib, et al. 2019b. Monolayer $Ti_3C_2T_x$ as an Effective Co-Catalyst for Enhanced Photocatalytic Hydrogen Production over TiO_2. *ACS Applied Energy Materials* 2 (7):4640–4651.

Su, Tongming, Rui Peng, Zachary D. Hood, et al. 2018. One-Step Synthesis of Nb_2O_5/C/ Nb_2C (MXene) Composites and Their Use as Photocatalysts for Hydrogen Evolution. *ChemSusChem* 11 (4):688–699.

Su, Tongming, Zuzeng Qin, Hong-bing Ji, Yue-xiu Jiang, and Guan Huang. 2016. Recent Advances in the Photocatalytic Reduction of Carbon Dioxide. *Environmental Chemistry Letters* 14 (1):99–112.

Su, Tongming, Zuzeng Qin, Hongbing Ji, and Zili Wu. 2019. An Overview of Photocatalysis Facilitated by 2D Heterojunctions. *Nanotechnology* 30 (50):502002.

Su, Tongming, Qian Shao, Zuzeng Qin, Zhanhu Guo, and Zili Wu. 2018. Role of Interfaces in Two-Dimensional Photocatalyst for Water Splitting. *ACS Catalysis* 8 (3):2253–2276.

Wang, Wuyou, Zachary D. Hood, Xuanyu Zhang, et al. 2020. Construction of 2D $BiVO_4$– CdS–$Ti_3C_2T_x$ Heterostructures for Enhanced Photo-Redox Activities. *ChemCatChem* 12 (13):3496–3503.

Wu, Shuai, Hongtao Yu, Shuo Chen, and Xie Quan. 2020. Enhanced Photocatalytic H_2O_2 Production over Carbon Nitride by Doping and Defect Engineering. *ACS Catalysis* 10 (24):14380–14389.

Yang, Xianglong, Xiao Xu, Jian Wang, et al. 2021. Insights into the Surface/Interface Modifications of Bi_2MoO_6: Feasible Strategies and Photocatalytic Applications. *Solar RRL* 5 (2):2000442.

Zhang, Wenhao, Abdul Rahman Mohamed, and Wee-Jun Ong. 2020. Z-Scheme Photocatalytic Systems for Carbon Dioxide Reduction: Where Are We Now? *Angewandte Chemie International Edition* 59 (51):22894–22915.

2 Preparation Method for MXene

Qingqing Li and Zuzeng Qin
School of Chemistry and Chemical Engineering, Guangxi University

CONTENTS

2.1 INTRODUCTION

In recent years, two-dimensional materials have been widely used in the fields of photocatalysts, sensors, electrochemistry, and capacitors. The most familiar two-dimensional material is graphene, which has both ceramic and metal properties and has been widely studied. In addition to the well-known two-dimensional materials such as graphene, MXene, a newly discovered two-dimensional material, has been widely studied due to its excellent properties. MXene phase is a two-dimensional material obtained by etching out the "A" atoms in the 3D material MAX phase. To understand the preparation of MXene, it is necessary to understand the MAX phase and its preparation principle, as well as the preparation method and process.

2.2 OVERVIEW OF MAX

MAX phase is the precursor for 2D MXene synthesis. As shown in Figure 2.1a, MAX is a hexagonal-layered transition metal carbide and nitride. Generally, $M_{n+1}AX_n$ (n = 1, 2, or 3), where M, A, and X represent transition metals, elements IIA or IVA, and C and/or N atoms, respectively. The earliest MAX material dates back to the 1960s, when it was first proposed by Prof. Nowotny at the University of

DOI: 10.1201/9781003156963-2

5

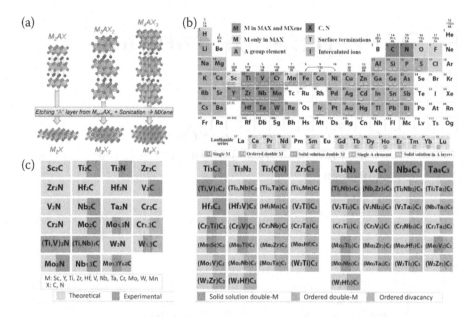

FIGURE 2.1 (a) MAX phase and the derived original MXene framework (Naguib et al. 2014), (b) The periodic table shows the composition of the MAX phase and MXene, (c) Composition of MXene as reported to date (Gogotsi and Anasori 2019).

Vienna, who called it the H-phase, or M_2BC (Jeitschko and Nowotny 1967). Since then, compounds with similar structures, such as Ti_3SiC_2, Ti_3GeC_2, and Ti_3AlC_2 (Pietzka and Schuster 1994), have been discovered one after another. Until 2000, Prof. Barsoum conducted a comprehensive review of these materials and collectively referred to these materials as MAX phase (Barsoum 2000).

MAX phase ceramics belong to a hexagonal crystal system, and the spatial group is P63/ MMC. Their excellent performance comes from their unique crystal structure. The tightly packed M atomic layer is stacked alternately with A atomic layer, and X atoms occupy the octahedral space composed of M atoms. It can also be described as the alternating arrangement of the M_6X octahedron and the A atomic layer along the C axis. As shown in Figure 2.1a, n values in the general formula "$M_{n+1}AX_n$" are different, and the number of MX slice layers interspersed between each A atomic layer is different. The M_2AX (n = 1) crystal structure contains two MX slice layers, M_3AX_2 (n = 2) crystal structure has three MX slice layers, and M_4AX_3 (n = 3) crystal structure contains four MX slice layers. The atomic arrangements of V_2AlC (Chenxu et al. 2016), Ti_3SiC_2 (Istomin et al. 2017), and β-Ta_4AlC_3 (Deng, Fan, and Lu 2009) in the direction of ($1\bar{2}10$) are shown in Figure 2.1a. In V_2AlC, every two rows of bright V atomic layers are separated by a dark Al atomic layer. And in Ti_3SiC_2, every three rows of bright Ti atomic layers are separated by a dark Al atomic layer, and every four rows of bright Ta atomic layers β-TA_4AlC_3 are separated by a dark Al atomic layer.

Recent literature indicated more than 60 MAX phases with ordered dual-transition metal structures, and Ti_3AlC_2 is the most widely studied (Barsoum 2013). Based on a variety of transition metals and their combinations with C and N elements, hundreds of possible phase MAX computations have been conducted (Figure 2.1b), and over 30 different MXenes have been identified (Figure 2.1b), as shown (Gogotsi and Anasori 2019).

In general, the MAX prophase was usually prepared at above 1500 °C, as well as by the need to test dozens of hours of synthesis condition. However, the researchers tried and practiced very hard and explored the more moderate synthesis conditions, higher-purity sample preparation methods, such as pressureless sintering, mechanical activation method, self-propagating high-temperature synthesis, hot-pressing method, and static method. The study on Ti_3AlC_2 has been applied to industrial mass production, which is one of the reasons why more and more researchers in MXene have joined the investigation.

2.3 SYNTHESIS OF MXENE

The synthesis method of MXene has generated extensive interest, and many methods have been developed, which can be summarized into two main strategies: top-down (for example, etching from MAX or non-Max precursors) and bottom-up (for example, chemical vapor deposition (CVD)).

2.3.1 PREPARATION OF MXENE BY THE TOP-DOWN ROUTE

2.3.1.1 Preparation of MXene by Etching MAX Phase with HF

Based on the weak interlayer van der Waals forces between the layers, most of the previous two-dimensional materials, such as graphene, metal dihalides, and black phosphorus, usually will be prepared by simple mechanical stripping (Dhanabalan et al. 2017). However, compared with other layered materials, such as graphite and transition metal dichalcogenides (TMDs), in which the structure is held together by weak van der Waals interactions, the binding force between the layers in the MAX phase is too strong to be destroyed by stripping or any similar mechanical means. Furthermore, the M-A bond in the MAX phase is more chemically active than the M-X bond. Therefore, it becomes a breakthrough point for preparing layered two-dimensional material MXene, and selective etching of element A layer in MAX becomes possible. Naguib (Naguib et al. 2011) reported the first type of Ti_3C_2 MXene by etching the MAX using HF. The Ti_3AlC_2 powder was immersed in HF solution of 50 wt% at room temperature for etching 2 h (Figure 2.2). The resulting suspension was washed several times with deionized water and centrifuged to separate the powder, as shown in Figure 2.3e. Its etching principle shows in Eqs. (2.1)–(2.3).

$$Ti_3AlC_2 + 3HF = AlF_3 + 3/2H_2 + Ti_3C_2 \tag{2.1}$$

$$Ti_3C_2 + H_2O = Ti_3C_2(OH)_2 + H_2 \tag{2.2}$$

$$Ti_3C_2 + 2HF = Ti_3C_2F_2 + H_2 \tag{2.3}$$

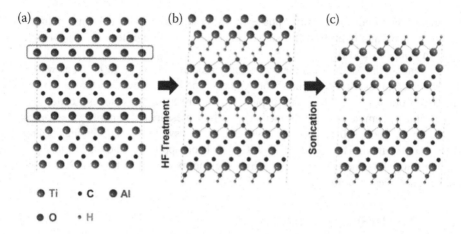

FIGURE 2.2 Schematic of the exfoliation process for Ti_3AlC_2. (a) Ti_3AlC_2 structure, (b) Al atoms replaced by OH after reaction with HF, (c) Breakage of the hydrogen bonds and separation of nanosheets after sonication in methanol (Naguib et al. 2011).

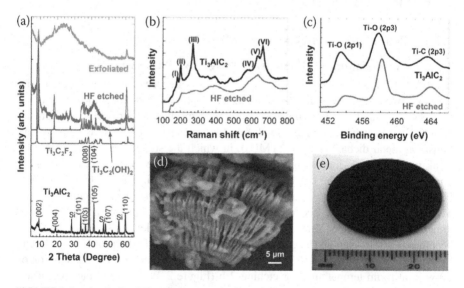

FIGURE 2.3 Analysis of Ti_3AlC_2 before and after exfoliation. (a) XRD pattern for Ti_3AlC_2 before HF treatment, simulated XRD patterns of $Ti_3C_2F_2$ and $Ti_3C_2(OH)_2$, measured XRD patterns of Ti_3AlC_2 after HF treatment, and exfoliated nanosheets produced by sonication. (b) Raman spectra of Ti_3AlC_2 before and after HF treatment. (c) XPS spectra of Ti_3AlC_2 before and after HF treatment. (d) SEM image of a sample after HF treatment. (e) Cold-pressed 25 mm disk of etched and exfoliated material after HF treatment (Naguib et al. 2011).

Eq. (2.1) predominates the whole reaction, resulting in Ti_3C_2, which then reacts with H_2O and HF molecules to form OH or F surface groups on MXene.

The XRD pattern shows that most of the nonbase peak of Ti_3AlC_2 has disappeared, especially the strongest peak at $2\theta = 39°$ (the middle curve in Figure 2.3a).

On the other hand, compared with the position before treatment, the peak of (00 1) becomes wider, and the strength decreases and moves to a lower angle, indicating Al has been dissolved by the etch reaction, resulting in the increase of the formation of multilayer of two-dimensional structure. The strongest diffraction peak of $Ti_3C_2(OH)_2$ disappears at $2\theta = 39$, and the simulated XRD pattern of $Ti_3C_2(OH)_2$ is in good agreement with the experimental results, which provides strong evidence for the formation of $Ti_3C_2(OH)_2$. The existence of OH groups after treatment can be confirmed by Fourier transform infrared (FTIR) spectrum, which fully proves the existence of Eqs. (2.2) and (2.3). On the other hand, it can also be explained from the perspective of stability: the Ti atoms exposed on the MXene surface are unstable in the air and should be satisfied by appropriate ligands; at the same time, there are a large number of OH and F ions in the etching solution; therefore, there will be OH and F ions on the surface to stablize the Ti atoms. The layered organ-like $Ti_3C_2(OH)_2$ obtained by etching is shown in Figure 2.3d. To follow the conventions used in the graphene/nanotube field, these loosely packed particles are called multilayer or ML-MXenes. When the number of stacked layers is less than five, they are referred to as low-layer MXene (FL-MXene). Since a variety of surface terminators are possible, a common labeling scheme is required. Therefore, the general formula of MXene is usually expressed as: $M_{n+1}X_nT_x$, where T is the functional group terminated on the surface (OH, F, O, and H).

It is worth noting that if the MAX phase is completely converted to MXene, all the peaks (000 1) in the XRD pattern will weaken or disappear (Ghidiu, Naguib, et al. 2014), especially if the M_2X structure is thinner. In addition, the (000 1) peak should not only be widened but also should be down-shifted to a smaller angle, indicating the C-lattice parameter is larger. However, in impure transformation, the MAX phase and the diffraction peak of Mxene will exist simultaneously (Figure 2.4).

Generally, the movement of the stronger and smaller angle of the (001) peak in the XRD pattern is used as a symbol for the increased layer spacing, and the disappearance of the peak at $2\theta = 39°$ is used as a symbol of the disappearance of Al in the MAX phase to judge whether the Al successfully etched to two-dimensional

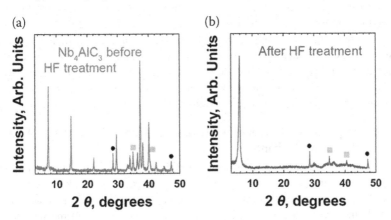

FIGURE 2.4 XRD patterns of Nb_4AlC_3 (a) before and (b) after HF treatment (Ghidiu, Naguib, et al. 2014).

Mxene from the MAX. At the same time, since the XRD peak strength decreases with the increasing stripping degree, namely, the number of layers of Mxene decreases, the XRD cannot quantify the proportion of unreacted MAX phase in the sample alone. To solve this problem, energy-dispersive spectroscopy (EDS) can be used to quantify the atomic ratio of A:M in the Mxene. However, the problem in this method is that the element A may also exist in the Mxene sample in the form of salt-containing A, due to incomplete removal in the washing process; this results in a MAX phase ratio tending to be larger than the actual proportion by EDS testing.

Since then, many new MXene, Ti_2C, V_2C, NB_2C, Nb_4C_3 and Ta_4C_3 using HF etching MAX have been synthesized (Liu, Mao-Cheng, et al. 2021; Gao, Lingfeng, et al. 2021 and Ghidiu, M, et al. 2014). In addition, Mxene with two randomly distributed transition elements, such as $(V_{0.5}Cr_{0.5})_3C_2$, $(Ti_{0.5}Nb_{0.5})_2C$, and $(Nb_{0.8}Ti_{0.2})_4C_3$ (Yang et al. 2016), were also synthesized. Anasori (Anasori et al. 2015) firstly synthesized the ordered double-M elements Mxene (Mo_2TiC_2 and $Mo_2Ti_2C_3$) and sandwiched $Mo_2Ti_2C_3$ by HF, which further extending the 2D material series. This method is mostly used in the MAX etching of "A = Al", and only a very few non-Al elements MAX are etched into Mxene. Part of Mxene prepared by HF etching is shown in Table 2.1.

Although element Al in the MAX phase is all etched, the etching concentration, temperature, time, and other requirements are very different from Table 2.1. These works show that the structure, atomic bond number, and particle size of MAX have great influence on the etching conditions. When the number of M atoms increases, the M-A bond energy increases, and a stronger etching force is needed to destroy the M-Al metal bond; therefore, the concentration of fluoride ion in the etching-agent solution and the etching time should both increase. Larger n requires longer etching time and a greater HF concentration. For example, the etching time of Nb_2AlC (90 h) is three times as much as that of Ti_2AlC (24 h). HF concentration at room temperature is approximately 50%. As the etching temperature increased to 55 °C, the etching time of Nb_2AlC decreased from 90 h to 40 h. In general, Mxene with a larger n in $M_{n+1}X_n$ requires a higher HF concentration or longer etching time to achieve complete conversion. For example, under the same etching conditions, the etching time of $Mo_2Ti_2AlC_3$ (96 h) is twice as much as that of Mo_2TiAlC_2 (48 h). However, the longer the etching time, the more favorable it will be. After immersing V_4AlC_3 powder into a 48% HF aqueous solution for 10 days at room temperature, The $V_4C_3T_x$ quantum dots will be formed, as shown in Figure 2.3d in literature [29], which may cause the etching of V atoms by HF or the etching defects in nanometer flakes (Zhao et al. 2020).

2.3.1.2 Preparation of Mxene by Etching MAX Phase with Mixed Acid and Other Etchants

From the above conclusion, the etching conditions are very important for synthesizing Mxene from the MAX phase. In insoluble Mxene, generally speaking, the higher the reaction conditions are, the more favorable the etching reaction will be. The higher the acid concentration is, the shorter the reaction time will be, and the lower the acid concentration is, the longer the reaction time and higher the reaction temperature will be. HF etching method has the problem that HF toxicity

TABLE 2.1

Preparation of MXene Synthesis by Etching MAX with HF

MAX	Mxene	HF (con)	Time (h/d)	T (°C)	Con (%)	dMAX	dMXene	Ref
V_2AlC	V_2CT_x	50	8	RT	55	13.13	23.96	(Naguib et al. 2013)
V_2AlC	V_2CT_x	50	90	RT	60	13.13	19.73	(Wang, Yuan, et al. 2016)
V_2AlC	V_2CT_x	40	7 d	Rt	68	–	23.7	(VahidMohammadi et al. 2017)
V_2AlC	V_2CT_x	50	92	RT	60	13.88	22.34	
Nb_2AlC	Nb_2CT_x	50	90	RT	100	–	–	(Naguib et al. 2013)
Ti_2AlN	Ti_2NT_x	5	24	RT	NA	13.79	14.88	(Soundiraraju and George 2017)
$TiNbAlC$	$(Ti_{0.5},Nb_{0.5})_2CT_x$	50	28	RT	80	1.773	2.426	(Naguib et al. 2012)
$(V_{0.5}Cr_{0.5})_3AlC_2$	$(V_{0.5}Cr_{0.5})_3C_2T_x$	50	69	RT	NA	–	–	(Seredych et al. 2019)
Mo_2Ga_2C	Mo_2CT_x	50	120	50	–	–	–	(Seredych et al. 2019)
Nb_2AlC	Nb_2CT_x	50	48	50	80	–	–	(Zhu et al. 2019)
Ti_2AlC	Ti_2CT_x	10	10	RT	–	–	9.7	(Zhao et al. 2017)
Nb_4AlC_3	$Nb_4C_3T_x$	49	140	RT	–	9.3	29.63	(Anayee et al. 2020)
Ti_3AlC_2	$Ti_3C_2T_x$	5	15 h	40	–	22.719	36.0	
V_4AlC_3	$V_4C_3T_x$	40	165	RT	–	28.2	25.8	(Tran et al. 2018)
Mo_4VAlC_4	$Mo_4VC_4T_x$	48–50	8d	50	100	18.6	34.6	(Deysher et al. 2020)
Mo_2TiAlC_2	$Mo_2TiC_2T_x$	48–51	48	RT	–	23.6	19.4	(Anasori et al. 2015)
$Mo_2Ti_2AlC_3$	$Mo_2Ti_2C_3T_x$	48–51	90	55	100	13.8738	25.3	(Anasori et al. 2015)
$(Mo_{2/3}Sc_{1/3})_2AlC$	$Mo_{1.33}CT_x$	48	24	RT	100	13.9736	32.53	(Tao et al. 2017)
$(Nb_{2/3}Sc_{1/3})_2AlC$	$Nb_{1.33}CT_x$	48	30	RT	–	27.73	32.70	(Halim et al. 2018)
$Zr_3Al_3C_5$	$Zr_3C_2T_x$	50	72	RT	65	31.88	–	(Zhou et al. 2016)
$Hf_3[Al(Si)]_4C$	$Hf_3C_2T_x$	35	60	RT	–	–		(Zhou et al. 2017)
$(Nb_{0.8},Ti_{0.2})_4AlC_3$	$(Nb_{0.8},Ti_{0.2})_4C_3T_x$	50	96	50				(Yang et al. 2016)

leads to a high-risk coefficient, which needs further development. Ghidiu (Ghidiu, Lukatskaya, et al. 2014) used a safer mixed solution of hydrochloric acid (HCl) and lithium fluoride (LiF), and soaked Ti_3AlC_2 powder in the above mixed solution of LiF and HCl at 35 °C for 24 h to obtain $Ti_3C_2T_x$. Compared with HF acid etching Mxene, fewer defects were found in the latter, indicating this method is efficient and safer; it has been be used in the synthesis of other Mxene, including Mo_2CT_x, Ti_3CNT_x, $(Nb_{0.8}ZR_{0.24})C_3T_x$, Mo_2CT_x. Wang (Wang, Zhang, et al. 2016) demonstrated a safe and straightforward hydrothermal method to prepare Ti_3C_2 Mxene by soaking Ti_3AlC_2 powder in ammonium fluoride (NH_4F) solution. In fact, since H and F ions are present in the solution, the principle of this method is the same as that of HF etching. When HCl/LiF or NH_4F solution is used, HF can be generated in situ by the following reaction, as show in Eqs. (2.4) and (2.5).

$$LiF + HCl = HF + LiCl \qquad (2.4)$$

$$NH_4F + H_2O = NH_3H_2O + HF \qquad (2.5)$$

The HF generated in situ was further reacted with the MAX in the same principle of Eqs. (2.1)–(2.3), and MXene was synthesized by selective etching of Al. The method of etching acid and hydrofluorite (i.e., LiF, KF, and FeF) mixed solution has been used to synthesize many different MXene families. Kvashina (Kvashina et al. 2020) found the HCl and NH_4F etching was mainly influenced by the hydrofluorate amount and the treatment time of MAX phase by containing the etching agent in the solution. Meanwhile, the study showed that the optimal etching result was obtained at a NH_4F concentration of 3 M and a treatment time of 160 h. Unlike MXene obtained by concentrated HF etching, the MXene flakes show negligible nanometer defects using HCl/LiF due to relatively mild acid and fluoride-etching conditions.

In addition, the presence of Li ions is conducive to the selection of Al etching. When the ratio of MAX to LiF doubles from 1:5 to 1:7.5, the scanning electron microscope (SEM) image showed most of the flakes produced by the former are 200–500 nm in diameter; in contrast, the latter produced is much larger MXene flakes with a diameter of 4–15 m (Figure 2.5), which looks uniform and have the same thickness, and the quality of the synthesized MXene is greatly improved, and

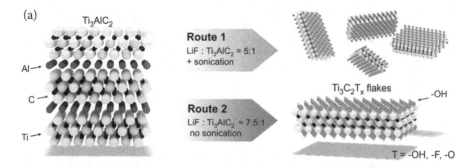

FIGURE 2.5 Etching process of Ti_3AlC_2 (Lipatov et al. 2016).

the mechanical strength stability and other properties of MXene are also improved. Moreover, the multilayer $Ti_3C_2T_x$ can be layered into a single layer by manual shaking, without the ultrasonic treatment of cutting the slices into small pieces (Lipatov et al. 2016). Furthermore, when the solution contains NH_4^+, the MXene layer can be inserted by these ions due to the presence of Li^+ and NH_4^+ cations, which can further expand the interlayer spacing and weaken the interaction between the MXene layers, facilitating the further stratification of MXene. So far, many MXene, including Ti_2CT_x (Liu, Zhou, Chen, et al. 2017), Mo_2CT_x(Guo et al. 2020), V_2CTx(Liu, Zhou, Wang, et al. 2017), $Ti_3C_2T_x$(Ghidiu, Lukatskaya, et al. 2014) and $(Nb_{0.8}Zr_{0.2})_4C_3T_x$ (Yang et al. 2016), have been successfully prepared by using the mixed solution of fluorine ions in acid and salt.

Mixed acids reduce the risk of the etching synthesis of MXene; however, since the etching reaction is essentially constant, HF will be generated in situ. Therefore, in the etching process, harmful HF gas will be released from the mixed solution. Furthermore, previous density-functional theory (DFT) calculations predicted the surface functionalization of F would prevent the transport of Li^+ ions and reduce the electrochemical capacity of Li^+ storage (Tang, Zhou, and Shen 2012; Enyashin and Ivanovskii 2013). To solve this problem, Ti_3C_2 MXene was obtained by using a safe and effective method and mild hydrogen-fluoride etchers, such as NH_4HF_2, $NaHF_2$, and KHF_2. Aihu Feng (Feng et al. 2017) found that weakly acidic and environmentally friendly ammonium fluoride hydroxide (NH_4HF_2) could synthesize MXenes, and they do not produce harmful gases. The reaction mechanism is slightly different from that of concentrated HF solution, which can be summarized as Eqs. (2.6)–(2.8).

$$2Ti_3AlC_2 + 3YAF_2 = Y_3Al_3 + AlF_3 + 3/2H_2 + 2Ti_3C_2 \qquad (2.6)$$

$$AlF_3 + aH_2O = AlF_3 \cdot aH_2O \qquad (2.7)$$

$$Ti_3C_2 + YHF_2 + H_2O = Ti_3C_2F_b(OH)_cY_d \qquad (2.8)$$

where, $Ti_3C_2F_b(OH)_cY_d$ is the chemical formula of Ti_3C_2 MXene with various surface terminals; the inserted Y cations, e.g., NH_4^+, Na^+, or K^+, will further expand the crystal plane spacing while maintaining the 2D wafer structure of the resulting MXene.

The Mxenes synthesis by other methods are listing in Table 2.2.

F has a strong bonding force. F-containing solution can be formed from the classic MAX phase-selective etching A MXene; however, this method is suitable for removing the alkaline/F amphoteric elements, and it is not suitable for removing the acid element, i.e., sulfur, and phosphorus in the V_2PC Zr_2SC. That is why most of the current MXene successfully synthesized from the A MAX phase consists of Al, Si or Ga atom. At the same time, the MXene synthesis takes F ion as the surface-sealing terminal and functional ion, and F is a corrosive material, which is not only harmful to the environment, but also corrosize to the battery, which reduces the material performance (for example, the capacitance) (Anasori, Lukatskaya, and Gogotsi 2017).

TABLE 2.2

Summary of MXene Synthesis by Other Methods

MAX	MXene	Etching Agent	Time (h)	T (°C)	Con (%)	Ref
Ti_3AlC_2	$Ti_3C_2T_x$	LiF + HCl	45	40	100	(Ghidiu, Lukatskaya, et al. 2014)
Ti_2AlN	Ti_2NT_x	KF + HCl	3	RT	87	(Soundiraraju and George 2017)
Mo_2Ga_2C	Mo_2CT_x	LiF + HCl	16 d	RT	100	(Halim et al. 2016)
V_2AlC	V_2C	NaF + HCl	72	90	90	(Liu, Zhou, Wang, et al. 2017)
Ti_3AlCN	Ti_3CNT_x	LiF + HCl	30	RT	–	(Du et al. 2017)
$(Nb_{0.8}Zr_{0.24})AlC_3$	$(Nb_{0.8}Zr_{0.24})C_3T_x$	LiF + HCl	168	50	42	(Yang et al. 2016)
Mo_2Ga_2C	Mo_2CT_x	LiF + HCl	24	140	100	(Guo et al. 2020)
TiVAlC	TiVCTx	LiF + HCl	60	55	–	(Yazdanparast et al. 2020)
$(V_xTi_{1-x})_2AlC$	$(V_xTi_{1-x})_2C$	LiF + HCl	24	90	–	(Wang et al. 2019)
Ti_3AlC_2	$Ti_3C_2T_x$	FeF_3 + HCl	50	60	100	(Wang et al. 2017)
Ti_3AlC_2	$Ti_3C_2T_x$	NH_4HF_2	11	RT	–	(Halim et al. 2014)
Ti_3AlC_2	$Ti_3C_2T_x$	NH_4F	36	180	100	(Wang, Zhang, et al. 2016)
Ti_4AlN_3	$Ti_4N_3T_x$	LiF + NaF + KF	30	550	–	(Urbankowski et al. 2016)
Ti_3SiC_2	$Ti_3C_2T_x$	35% H_2O_2 + 30% HF	24	80	–	(Alhabeb et al. 2018)

Therefore, to etch other MAX phases to extend the MXene family, new synthetic MXene methods need to be studied.

In addition to the conventional F-containing aqueous solution treatment, MXene can also be produced by etching its parent MAX phase at high temperatures. Xuan (Xuan et al. 2016) oxidized tetramethylammonium with organic-base hydrogen based on the amphoteric property, which states that Al can react with acid and base (TMAOH), as an etching agent; the etching steps are shown in Figure 2.6. The XRD shows the layer spacing extending from 0.92 to 1.50 nm, indicating the successful insertion of the TMA ions, and the phase transition is almost complete at 24 h. However, since the Ti_3AlC_2 surface exhibit a high tendency to be passivated by the thin oxide layer, very thin HF pretreatment is needed to remove the passivated layer, and HF and TMAOH are respectively added in two steps for the etching reaction. And the reaction principle of this method is different from that of MXene consuming "Al" in the HF acid-etching synthesis. The reaction between Al and TMAOH results in the hydrolysis of Al to $Al(OH)_4^-$. Due to its negative charge, $Al(OH)_4^-$ is immediately bonded to the titanium surface through an O-TI bond,

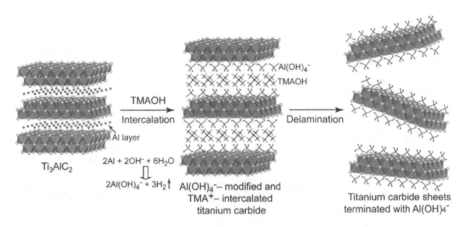

FIGURE 2.6 Schematic illustration showing the intercalation and delamination process (Xuan et al. 2016).

resulting in the filling of $Al(OH)_4^-$ and TMA^+ between the Ti_3C_2 layers. As shown in Figure 2.6, in this structure, an aluminum atom will be replaced with an $A(OH)_4^-$ rather than etched into the traditional MXene structure (Xuan et al. 2016).

Thermodynamically, Al can be reacted with OH^- in The MAX phase Ti_3AlC_2; however, why is it difficult to etch MXene? And even when a reaction occurs, why is the etching of Al atoms largely limited to the particles' surface? (Xie et al. 2014). The key issue is the kinetic, where some oxide/hydroxide layers, such as Al_2O_3, form on the surface of Ti_3AlC_2. However, the yield of Ti_3C_2Tx obtained by continuous sodium hydroxide (NaOH) treatment was still very low, even if the oxide coating was removed by a 10 wt% HF aqueous solution. The formation of new Al (oxide) hydroxyl layers with alkali exposure, such as $Al(OH)_3$ and $AlO(OH)$, may further hinder the required Al extraction (Xuan et al. 2016). ANaOH-assisted hydrothermal method was recently invented to prepare Ti_3C_2Tx (T = OH, O) by etching Ti_3AlC_2. In the synthesis of $Ti_3C_2T_x$, the temperature was the dominant key factor, while NaOH concentration affected the purity of MXene obtained. When the temperature and the alkali concentration are both low, the prepared samples have higher water content, which promotes the oxidation of Ti and produces Na-Ti-O compounds, such as $Na_2Ti_3O_7$ and $Na_2Ti_5O_{11}$. In a 27.5 M NaOH solution, Al was successfully selectively removed from Ti_3AlC_2 at 270 °C to obtain OH with a purity of 92 wt% and O terminal multilayer $Ti_3C_2T_x$ (Li et al. 2018). For the first time, MXene was successfully prepared by using only alkali solution and MAX etching. As for the synthesis of nitride MXene, there have been various reports of preparation attempts, but no suitable MXene has been formed.

Based on the first principle theory calculation, Shein and Ivanovskii (Shein and Ivanovskii 2012) proposed that the cohesive energy of Ti-N bond is lower than that of Ti-C bond, which leads to a low stability and a high solubility of $M_{n+1}N_n$ structure in HF solution. On the other had, the formation energy of Ti-N bond is much higher than that of Ti-C bond, which indicates that more energy is needed to extract Al element from $Ti_{n+1}Al_n$ phase. Therefore, it is difficult to etch $Ti_{n+1}N_nT_x$

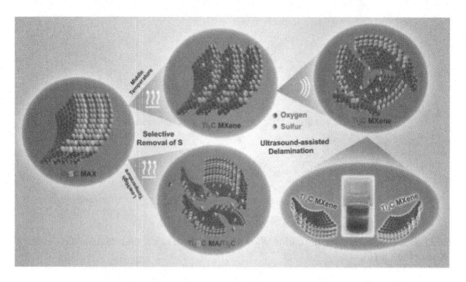

FIGURE 2.7 Schematic illustration of the fabrication of 2D Ti₂C MXene via an effective thermal-reduction approach (Mei et al. 2020).

MXene with hydrofluoric acid (Naguib et al. 2014). Until 2016, Patrick Urbankowski (Urbankowski et al. 2016) successfully prepared multilayer MXene $Ti_4N_3T_x$ by heating Ti_4AlN_3 and fluoride salt mixture at 550 °C for 30 min. Since then, Ti_2NT_x has also been successfully prepared by adding a mixture of KF and HCl (Soundiraraju and George 2017). In addition, an innovative thermal-reduction strategy (Mei et al. 2020), as shown in Figure 2.7, was proposed for the preparation of MXene from the corresponding sulfur-containing MAX phase in which the weakly bonded S atoms reacted with hydrogen to form a volatile gas, leaving a two-dimensional grapheme-like Ti_2C nanosheet, successfully preparing the two-dimensional MXene structure from the S-based MAX for the first time. Among them, temperature plays a decisive role; the optimal synthesis temperature of MXene is 800 °C. Until now, the production of 2D MXene by extracting relatively inert sulfur from the corresponding sulfur-containing MAX skeleton had not been reported; the reason is only amorphous carbon/sulfur sheets could be obtained if the usual fluorine-containing acid aqueous solution was used. This controlled thermal-reduction strategy not only provides a brand-new etching method for other MAX of the etching for the family, but it is also easy to expand, avoids the use of harmful acid, and has a great potential in industrial production of the future. However, there is a certain danger from the use of hydrogen and the higher reaction temperature, and the generated etching S by the product need for further processing.

2.3.1.3 Preparation of Monolayer and Few-Layer MXenes

After HF etching, the resulting multilayer films are usually washed with water several times to remove the synthetic by-products, such as aluminum fluoride (AlF_3) (Naguib et al. 2011); prewashes with HCl or sulfuric acid (H_2SO_4) can also be used to help dissolve salts such as AlF_3 or LiF (Ghidiu et al. 2016; Urbankowski et al. 2016). Only in this way can the multilayer film be peeled off and a colloidal suspension consisting of one or several MXene layers obtained. The stripping technique depends on (1) the etching method and (2) the MXene composition.

After "A" is selectively etched by HF through etchant, multilayer MXene is obtained. Currently, there are two methods, mechanical stratification and inter-stratification, for the stratification of multilayer MXene. In the first method, the yield of single-layer MXene is few, and multilayer MXene has strong interlayer interaction, making it impossible to obtain large-scale layered MXene flakes by a simple mechanical stripping method. Most multilayer liquid exfoliation through molecular intercalations yields a high-yield monolayer of MXene, i.e., up to 20 mg MXene per milliliter of solution. Similar to clay, the multilayered structure can accommodate various cations and organic molecules (Ghidiu et al. 2016, 2017). The introduction of appropriate molecules can cause the expansion of interlaminar space and the simultaneous weakening of interlaminar interaction. Therefore, ultrasound and shaking can lead to the stripping of MXene and its dispersion in water or other solvents.

In general, intercalations include polar organic molecules and polar organic base molecules. Polar organic molecule dimethyl sulfoxide (DMSO) has been used to strip Ti_3C_2Z (Mashtalir et al. 2013) and $(Mo_{2/3}Ti_{1/3})_3C_2Tz$ (Anasori et al. 2015). Mashtalir (Mashtalir et al. 2013) inserted dimethyl sulfoxide (DMSO) between the Ti_3C_2 nanorods etched by HF and treated with ultrasonic in water, resulted in minimal or monolayer MXene. The (000l) peak in the XRD pattern moves to a lower angle of $2°$, demonstrating the expansion of the interlaminar space. The (0002) peak in the XRD pattern corresponds to half of the C lattice parameter and is hereafter referred to simply as the interval d between layers. Wu (Wu et al. 2017) obtained several layers of MXene nanosheets with high yield by a high-energy mechanical grinding method using DMSO as the solvent and embedding agent. The colloidal solution of delaminated MXene was very stable, and the Tyndall effect was observed when the red laser beam was irradiated. This method of layered DMSO as an embedder is widely used (Mashtalir et al. 2013; Elumalai et al. 2019; Rajavel et al. 2019; Liu et al. 2020; Mazhar et al. 2020), but DMSO is not effective for other MXene except for these two components.

In Figure 2.9, the blue spheres represent nitrogen atoms; red represents niobium; black represents carbon, and white represents hydrogen. MXene's surface terminations are not shown (Mashtalir et al. 2015).

DMSO auxiliary-stratification method, which only plays a role in $Ti_3C_2T_x$, cannot meet the actual requirements. The researchers found that other MXene layers, such as Ti_3CNT_x, V_2CT_X and Nb_2CT_x (Figure 2.8), could be easily stratified by ultrasound treatment after embedding tetrabutylammonium hydroxide (TBAOH) (Naguib et al. 2015), tetrapothylammonium hydroxide (TPAOH) (Lin et al. 2017), and iso-propylamine (i-PrA) (Mashtalir et al. 2015). Researchers believe that i-PrA can form ammonium cation $R-NH_3^+$ when mixed with water, and electrostatic assistance may be inserted between the Nb_2CT_X layer and the Nb_2CT_X layer. Secondly, the I-PRA molecule has a three-carbon alkyl tail, which may be small enough to overcome the steric hindrance during the insertion, but large enough to push the MXene layer aside, resulting in the reduced interlayer interaction and accompanying stratification, as shown in Figure 2.9 (Mashtalir et al. 2015). TBAOH was successfully used to remove

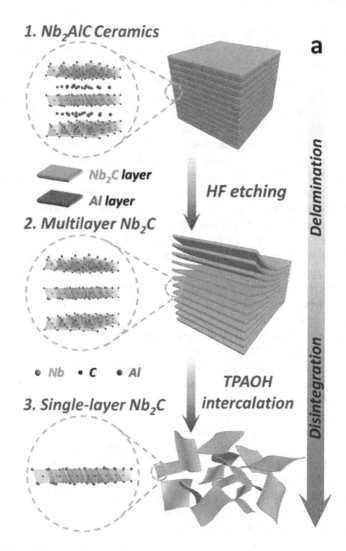

FIGURE 2.8 Schematic diagram for the fabrication of ultrathin Nb₂C nanosheets, including HF etching (delamination) and TPAOH intercalation (disintegration) (Lin et al. 2017).

many MXene compositions, for example, $(Mo_{2/3}Ti_{1/3})_3C_2T_z$ and $(Mo_{1/2}Ti_{1/2})_4C_3T_z$ (Anasori et al. 2016), $Ti_4N_3T_z$ (Urbankowski et al. 2016), Mo_2CT_z (Halim et al. 2016). However, TBAOH cannot strip $Ti_3C_2T_z$, while TMAOH can strip $Ti_3C_2T_z$ (Alhabeb et al. 2017).

Treating MAX with a more environmentally friendly LIF-HCL etching agent would facilitate the pre-embedding of Li^+ ions during etching, increasing the layer spacing of MXene and reducing the interaction between layers. This would make it easier for multilayer MXene to be stratified during ultrasonic processing; therefore, MXene slices with few or single layers could be obtained. Lif-HCl is one of the most widely used in situ HF etchants for the synthesis of high-quality MXene. In

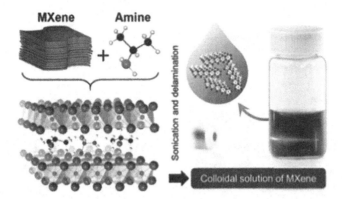

FIGURE 2.9 Schematic of Nb_2CT_x delamination process via isopropylamine intercalation. The photo on the right shows Tyndall scattering effect in a colloidal solution of $d-Nb_2CT_x$ flakes (Mashtalir et al. 2015).

addition, compared with distilled water, ethanol as a washing solution can significantly improve the stratification rate due to the weak interlayer force caused by larger interlayer molecules (Zhang et al. 2017).

2.3.2 PREPARATION OF MXENE BY THE BOTTOM-UP ROUTE

In addition to the top-down approach for MXene synthesis from MAX, the bottom-up approach is also applicable to the synthesis of MXene structures (Halim et al. 2014; Gogotsi 2015; Xu et al. 2015; Liu et al. 2016; Geng et al. 2017). Halim (Halim et al. 2014) reported the deposition of Ti, Al, and C element targets on an insulating sapphire substrate by a DC magnetron sputtering to prepare Ti_3AlC_2 MAX thin films. By selective etching of Al layer, Ti_3C_2 thin film of 1×1 cm^2 and 19 nm thick was obtained, with a transmittance of 90% in the range of visible light to infrared. In addition, both maximal phase (Mo_2GaC) and nonmaximal phase (Mo_2Ga_2C) films can be prepared by a DC magnetron sputtering method (Lai et al. 2017).

Furthermore, a simple etching process was used to prepare the extended Mo_2C thin films. The results further show that CVD could directly prepare ultra-thin MXene materials, which was a new method to prepare MXene-based materials. A 2D ultra-α-Mo_2C (3 nm) with a large surface area and high quality were prepared by the methane (used as carbon source) CVD method on the surface of Cu/Mo-alloy at temperatures above 1085 °C. By changing the experimental conditions, the size and thickness of Mo_2C crystal can be well controlled. The transverse size increases with the growth time, and the nucleation density increases with the growth temperature. All kinds of shapes, including the triangle, rectangle, hexagon, octagon, hexagon, rhombic, and dodecagon, could be obtained by this method, as shown in Figure 2.10; the obtained α-Mo_2C has a uniform thickness and a smooth surface, and, more importantly, in large area, no defects, confusion, or impurities were observed, indicating that CVD can synthesize high-quality Mo_2C two-dimensional structure. In addition, the crystals can be perfectly transferred to any target substrate

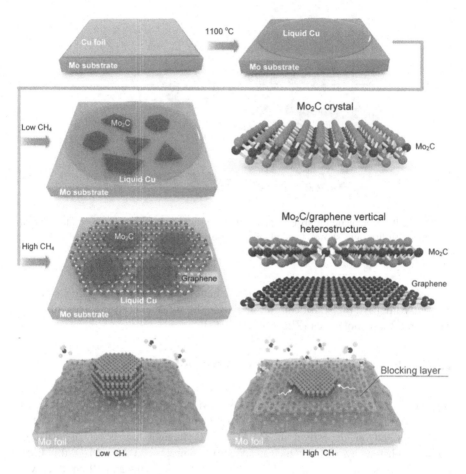

FIGURE 2.10 Schematic showing the growth of Mo_2C crystals under low and high flow rates of CH_4 (Geng et al. 2017).

by etching Cu using a 0.2 M $(NH_4)_2S_2O_8$ solution (Xu et al. 2015). Further study shows that the Mo atomic sublattice has a uniform hexagonal compact arrangement throughout the crystal without any boundary for all crystals. However, in addition to rectangular and octagonal crystals, the C atom sublattice consists of three carbon-atom equivalent octahedral-eccentric zigzag configurations, due to three or six domains with rotational symmetry and well-defined linear domain boundaries (Liu et al. 2016). Jia (Jia et al. 2017) further improved the method by using a MoO_2 nanometer sheet as a template and a Mo source to easily prepare an ultra-β-Mo_2C nanometer sheet through a rapid and scalable synthesis method.

Thick Mo_2C crystals with different shapes are randomly distributed on the Cu surface at low CH_4 concentrations. In contrast, at higher CH_4 concentrations, the growth of predominantly hexagonal-shaped, thin Mo_2C flakes on graphene is seen. The schematic shows the growth mechanism of Mo_2C at low and high CH_4, respectively. Note that at high CH_4 conditions, graphene serves as a blocking layer, cutting off the vertical supply of Mo, resulting in the drastically decreased thickness of Mo_2C grown on it.

It is proved that the morphology of Mo_2C crystal can be controlled by changing the concentration of CH_4, and a fractal Mo_2C crystal was obtained at a low CH_4 flow rate (0.1 sccm). With the increase of CH_4 concentration, the shape of the Mo_2C crystal changes from triangle to other polygons, which is caused by higher C supersaturation. When the CH_4 flow rate is 0.3, 0.4, and 0.6 sccm, the main shapes are rectangle, pentagon, and hexagon, respectively. In addition, by changing the thickness of Cu foil, the Mo_2C thickness of the crystal can vary from several hundred nanometers to several nanometers (Geng et al. 2016). Since then, Geng (Geng et al. 2017) reported the direct synthesis of 2D Mo_2C-graphene films by chemical vapor deposition (CVD) catalyzed by molten copper in one step; Mo_2C crystal morphology, to a great degree, is affected by the CH_4 flow: at low methane flux, growth is not uniform (Figure 2.10), includingtriangle, rectangle and pentagon, which is in line with the former research conclusion. In addition, graphene's presence influences the growth kinetics at higher methane fluxes, and the Mo_2C crystals grown on graphene are very uniform in shape and thickness. As shown in Figure 2.10, graphene can be used as a barrier layer, thus significantly reducing the thickness of the Mo_2C already grown. To synthesize two-dimensional MXene structures that meet the performance requirements, we must understand the influencing factors in the synthesis process. Turke (Turker et al. 2020) studied the influences of impurities, copper thickness, and graphene's presence on the morphology of growing Mo_2C crystal, and the results showed that: (1) Silicon atoms from quartz tube influenced the nucleation and growth behavior of Mo_2C, and to form thin crystals with large transverse size, it was necessary to cover the sample surface and prevent the deposition of silicon; (2) Compared with Mo_2C crystals formed on the surface of Cu, Mo_2C crystals formed on graphene are thinner and have fewer defects; these are caused by the presence of graphene, which acts as an additional diffusion barrier for Mo atoms from most copper; (3) The Cu layer acts as a valve to control the supply of Mo to form 2D Mo_2C crystals. To reduce the thickness and further control the shape of the crystals, the supply of Mo atoms can be reduced by increasing the thickness of the Cu layer. Understanding these relationships helps us synthesize high-quality, on-demand results.

Compared with two-dimensional materials prepared by different etching agents, MXenes prepared by CVD have fewer defects and no termination, which provides a platform for studying their inherent properties and domain boundary effects. To prepare other monolayer MXenes with multiple functions, further exploration of a bottom-up synthesis method is required, conducive to studying the inherent electronic and optical properties of MXenes.

ACKNOWLEDGMENTS

This work was supported by the National Natural Science Foundation of China (22078074, 21938001), Guangxi Natural Science Foundation (2019GXNSFAA-245006, 2020GXNSFDA297007, 2016GXNSFFA380015), Special funding for 'Guangxi Bagui Scholars', and Scientific Research Foundation for High-level Personnel from Guangxi University.

REFERENCES

Alhabeb, Mohamed, Kathleen Maleski, Babak Anasori, et al. 2017. Guidelines for Synthesis and Processing of Two-Dimensional Titanium Carbide (Ti_3C_2Tx MXene). *Chemistry of Materials* 29 (18):7633–7644.

Alhabeb, Mohamed, Kathleen Maleski, Tyler S. Mathis, et al. 2018. Selective Etching of Silicon from Ti_3SiC_2 (MAX) to Obtain 2D Titanium Carbide (MXene). *Angewandte Chemie International Edition* 57 (19):5444–5448.

Anasori, Babak, Maria R. Lukatskaya, and Yury Gogotsi. 2017. 2D Metal Carbides and Nitrides (MXenes) for Energy Storage. *Nature Reviews Materials* 2 (2):16098.

Anasori, Babak, Chenyang Shi, Eun Ju Moon, et al. 2016. Control of Electronic Properties of 2D Carbides (MXenes) by Manipulating Their Transition Metal Layers. *Nanoscale Horizons* 1 (3):227–234.

Anasori, Babak, Yu Xie, Majid Beidaghi, et al. 2015. Two-Dimensional, Ordered, Double Transition Metals Carbides (MXenes). *ACS Nano* 9 (10):9507–9516.

Anayee, Mark, Narendra Kurra, Mohamed Alhabeb, et al. 2020. Role of Acid Mixtures Etching on the Surface Chemistry and Sodium Ion Storage in Ti_3C_2Tx MXene. *Chemical Communications* 56 (45):6090–6093.

Barsoum, Michel W. 2000. The $M_{N+1}AX_N$ Phases: A New Class of Solids: Thermodynamically Stable Nanolaminates. *Progress in Solid State Chemistry* 28 (1):201–281.

Barsoum, Michel W. 2013. *MAX Phases: Properties of Machinable Ternary Carbides and Nitrides*. John Wiley & Sons.

Chenxu, Wang, Tengfei Yang, Jingren Xiao, et al. 2016. Structural Transitions Induced by Ion Irradiation in V_2AlC and Cr_2AlC. *Journal of the American Ceramic Society* 99:1769–1777.

Deng, X. H., B. B. Fan, and W. Lu. 2009. First-Principles Investigations on Elastic Properties of α- and β- Ta_4AlC_3. *Solid State Communications* 149 (11):441–444.

Deysher, Grayson, Christopher Eugene Shuck, Kanit Hantanasirisakul, et al. 2020. Synthesis of Mo_4VAlC_4 MAX Phase and Two-Dimensional Mo_4VC_4 MXene with Five Atomic Layers of Transition Metals. *ACS Nano* 14 (1):204–217.

Dhanabalan, Sathish Chander, Joice Sophia Ponraj, Zhinan Guo, Shaojuan Li, Qiaoliang Bao, and Han Zhang. 2017. Emerging Trends in Phosphorene Fabrication towards Next Generation Devices. *Advanced Science* 4 (6):1600305.

Du, Fei, Huan Tang, Limei Pan, et al. 2017. Environmental Friendly Scalable Production of Colloidal 2D Titanium Carbonitride MXene with Minimized Nanosheets Restacking for Excellent Cycle Life Lithium-Ion Batteries. *Electrochimica Acta* 235:690–699.

Elumalai, Satheeshkumar, Veerappan Mani, Nithiya Jeromiyas, Vinoth Kumar Ponnusamy, and Masahiro Yoshimura. 2019. A Composite Film Prepared from Titanium Carbide Ti_3C_2Tx (MXene) and Gold Nanoparticles for Voltammetric Determination of Uric Acid and Folic Acid. *Microchimica Acta* 187 (1):33.

Enyashin, Andrey N. and Alexander L. Ivanovskii. 2013. Structural and Electronic Properties and Stability of MXenes Ti_2C and Ti_3C_2 Functionalized by Methoxy Groups. *The Journal of Physical Chemistry C* 117 (26):13637–13643.

Feng, Aihu, Yun Yu, Yong Wang, et al. 2017. Two-Dimensional MXene Ti_3C_2 Produced by Exfoliation of Ti_3AlC_2. *Materials & Design* 114:161–166.

Gao, Lingfeng, Chunyang Ma, Songrui Wei, Artem V. Kuklin, Han Zhang, and Hans Ågren (2021). Applications of Few-Layer Nb_2C MXene: Narrow-Band Photodetectors and Femtosecond Mode-Locked Fiber Lasers. *ACS Nano*, 15, 954–965. 10.1021/acsnano.0c07608.

Geng, Dechao, Xiaoxu Zhao, Zhongxin Chen, et al. 2017. Direct Synthesis of Large-Area 2D Mo_2C on in Situ Grown Graphene. *Advanced Materials* 29 (35):1700072.

Geng, Dechao, Xiaoxu Zhao, Linjun Li, et al. 2016. Controlled Growth of Ultrathin Mo_2C Superconducting Crystals on Liquid Cu Surface. *2D Materials* 4 (1):011012.

Ghidiu, Michael, Joseph Halim, Sankalp Kota, David Bish, Yury Gogotsi, and Michel W. Barsoum. 2016. Ion-Exchange and Cation Solvation Reactions in Ti_3C_2 MXene. *Chemistry of Materials* 28 (10):3507–3514.

Ghidiu, Michael, Sankalp Kota, Joseph Halim, et al. 2017. Alkylammonium Cation Intercalation into Ti_3C_2 (MXene): Effects on Properties and Ion-Exchange Capacity Estimation. *Chemistry of Materials* 29 (3):1099–1106.

Ghidiu, Michael, Maria R. Lukatskaya, Meng-Qiang Zhao, Yury Gogotsi, and Michel W. Barsoum. 2014. Conductive Two-Dimensional Titanium Carbide 'Clay' with High Volumetric Capacitance. *Nature* 516 (7529):78–81.

Ghidiu, M., M. Naguib, C. Shi, et al. 2014. Synthesis and Characterization of Two-Dimensional Nb_4C_3 (MXene). *Chemical Communications* 50 (67):9517–9520.

Gogotsi, Yury. 2015. Transition Metal Carbides Go 2D. *Nature Materials* 14 (11):1079–1080.

Gogotsi, Yury and Babak Anasori. 2019. The Rise of MXenes. *ACS Nano* 13 (8):8491–8494.

Guo, Yitong, Sen Jin, Libo Wang, et al. 2020. Synthesis of Two-Dimensional Carbide Mo_2CTx MXene by Hydrothermal Etching with Fluorides and Its Thermal Stability. *Ceramics International* 46 (11, Part B):19550–19556.

Halim, Joseph, Sankalp Kota, Maria R. Lukatskaya, et al. 2016. Synthesis and Characterization of 2D Molybdenum Carbide (MXene). *Advanced Functional Materials* 26 (18):3118–3127.

Halim, Joseph, Maria R. Lukatskaya, Kevin M. Cook, et al. 2014. Transparent Conductive Two-Dimensional Titanium Carbide Epitaxial Thin Films. *Chemistry of Materials* 26 (7):2374–2381.

Halim, Joseph, J. Palisaitis, J. Lu, et al. 2018. Synthesis of Two-Dimensional $Nb_{1.33}C$ (MXene) with Randomly Distributed Vacancies by Etching of the Quaternary Solid Solution ($Nb_{2/3}Sc_{1/3}$)$_2$AlC MAX Phase. *ACS Applied Nano Materials* 1 (6):2455–2460.

Istomin, Pavel, Elena Istomina, Aleksandr Nadutkin, et al. 2017. Fabrication of Ti_3SiC_2 and Ti_4SiC_3 MAX Phase Ceramics Through Reduction of TiO_2 with SiC. *Ceramics International* 43 (18):16128–16135.

Jeitschko, W. and H. Nowotny. 1967. Die Kristallstruktur von Ti_3SiC_2—ein neuer komplexcarbid-typ. *Monatshefte Für Chemie Und Verwandte Teile Anderer Wissenschaften* 98 (2):329–337.

Jia, Jin, Tanli Xiong, Lili Zhao, et al. 2017. Ultrathin N-Doped $Mo2C$ Nanosheets with Exposed Active Sites as Efficient Electrocatalyst for Hydrogen Evolution Reactions. *ACS Nano* 11 (12):12509–12518.

Kvashina, T. S., N. F. Uvarov, M. A. Korchagin, Yu L. Krutskiy, and A. V. Ukhina. 2020. Synthesis of MXene Ti_3C_2 by Selective Etching of MAX-Phase Ti_3AlC_2. *Materials Today: Proceedings* 31:592–594.

Lai, Chung-Chuan, Hossein Fashandi, Jun Lu, et al. 2017. Phase Formation of Nanolaminated Mo_2AuC and $Mo_2(Au_{1-x}Ga_x)_2C$ by a Substitutional Reaction within Au-Capped Mo_2GaC and Mo_2Ga_2C Thin Films. *Nanoscale* 9 (45):17681–17687.

Li, Tengfei, Lulu Yao, Qinglei Liu, et al. 2018. Fluorine-Free Synthesis of High-Purity $Ti_3C_2T_x$ (T = OH, O) via Alkali Treatment. *Angewandte Chemie International Edition* 57 (21):6115–6119.

Lin, Han, Shanshan Gao, Chen Dai, Yu Chen, and Jianlin Shi. 2017. A Two-Dimensional Biodegradable Niobium Carbide (MXene) for Photothermal Tumor Eradication in NIR-I and NIR-II Biowindows. *Journal of the American Chemical Society* 139 (45):16235–16247.

Lipatov, Alexey, Mohamed Alhabeb, Maria R. Lukatskaya, Alex Boson, Yury Gogotsi, and Alexander Sinitskii. 2016. Effect of Synthesis on Quality, Electronic Properties and

Environmental Stability of Individual Monolayer Ti_3C_2 MXene FanFlakes. *Advanced Electronic Materials* 2 (12):1600255.

Liu, Mao-Cheng, Yu-Shan Zhang, Bin-Mei Zhang, Dong-Ting Zhang, Chen-Yang Tian, Ling-Bin Kong, and Yu-Xia Hu. 2021. Large Interlayer Spacing 2D Ta_4C_3 Matrix Supported 2D MoS_2 Nanosheets: A 3D Heterostructure Composite Towards High-Performance Sodium Ions Storage. *Renewable Energy* 169:573–581. 10.1016/j.renene.2021.01.051.

Liu, Shaogang, Libo Wang, Xiaolong Wang, Lu Liu, Aiguo Zhou, and Xinxin Cao. 2020. Preparation, Mechanical and Thermal Characteristics of d-Ti_3C_2/PVA Film. *Materials Today Communications* 22:100799.

Liu, Zhibo, Chuan Xu, Ning Kang, et al. 2016. Unique Domain Structure of Two-Dimensional α-Mo_2C Superconducting Crystals. *Nano Letters* 16 (7):4243–4250.

Liu, Fanfan, Aiguo Zhou, Jinfeng Chen, et al. 2017. Preparation of Ti_3C_2 and Ti_2C MXenes by Fluoride Salts Etching and Methane Adsorptive Properties. *Applied Surface Science* 416:781–789.

Liu, Fanfan, Jie Zhou, Shuwei Wang, et al. 2017. Preparation of High-Purity V_2C MXene and Electrochemical Properties as Li-Ion Batteries. *Journal of The Electrochemical Society* 164 (4):A709–A713.

Mashtalir, Olha, Maria R. Lukatskaya, Meng-Qiang Zhao, Michel W. Barsoum, and Yury Gogotsi. 2015. Amine-Assisted Delamination of Nb_2C MXene for Li-Ion Energy Storage Devices. *Advanced Materials* 27 (23):3501–3506.

Mashtalir, Olha, Michael Naguib, Vadym N. Mochalin, et al. 2013. Intercalation and Delamination of Layered Carbides and Carbonitrides. *Nature Communications* 4 (1):1716.

Mazhar, Sadaf, Awais Ali Qarni, Yasir Ul Haq, Zeeshan Ul Haq, and Imran Murtaza. 2020. Promising PVC/MXene Based Flexible Thin Film Nanocomposites with Excellent Dielectric, Thermal and Mechanical Properties. *Ceramics International* 46 (8, Part B):12593–12605.

Mei, Jun, Godwin A. Ayoko, Chunfeng Hu, and Ziqi Sun. 2020. Thermal Reduction of Sulfur-Containing MAX Phase for MXene Production. *Chemical Engineering Journal* 395:125111.

Naguib, Michael, Joseph Halim, Jun Lu, et al. 2013. New Two-Dimensional Niobium and Vanadium Carbides as Promising Materials for Li-Ion Batteries. *Journal of the American Chemical Society* 135 (43):15966–15969.

Naguib, Michael, Murat Kurtoglu, Volker Presser, et al. 2011. Two-Dimensional Nanocrystals Produced by Exfoliation of Ti_3AlC_2. *Advanced Materials* 23 (37):4248–4253.

Naguib, Michael, Olha Mashtalir, Joshua Carle, et al. 2012. Two-Dimensional Transition Metal Carbides. *ACS Nano* 6 (2):1322–1331.

Naguib, Michael, Vadym N. Mochalin, Michel W. Barsoum, and Yury Gogotsi. 2014. 25th Anniversary Article: MXenes: A New Family of Two-Dimensional Materials. *Advanced Materials* 26 (7):992–1005.

Naguib, Michael, Raymond R. Unocic, Beth L. Armstrong, and Jagjit Nanda. 2015. Large-Scale Delamination of Multi-Layers Transition Metal Carbides and Carbonitrides "MXenes". *Dalton Transactions* 44 (20):9353–9358.

Pietzka, M. A.and J. C. Schuster. 1994. Summary of Constitutional Data on the Aluminum-Carbon-Titanium System. *Journal of Phase Equilibria* 15 (4):392–400.

Rajavel, Krishnamoorthy, Shuyi Shen, Tao Ke, and Daohui Lin. 2019. Achieving High Bactericidal and Antibiofouling Activities of 2D Titanium Carbide (Ti_3C_2Tx) by Delamination and Intercalation. *2D Materials* 6 (3):035040.

Seredych, Mykola, Christopher Eugene Shuck, David Pinto, et al. 2019. High-Temperature Behavior and Surface Chemistry of Carbide MXenes Studied by Thermal Analysis. *Chemistry of Materials* 31 (9):3324–3332.

Shein, I. R., and A. L. Ivanovskii. 2012. Graphene-Like Titanium Carbides and Nitrides $Ti_{n+1}C_n$, $Ti_{n+1}N_n$ (n = 1, 2, and 3) from De-Intercalated MAX Phases: First-Principles Probing of Their Structural, Electronic Properties and Relative Stability. *Computational Materials Science* 65:104–114.

Soundiraraju, Bhuvaneswari, and Benny Kattikkanal George. 2017. Two-Dimensional Titanium Nitride (Ti_2N) MXene: Synthesis, Characterization, and Potential Application as Surface-Enhanced Raman Scattering Substrate. *ACS Nano* 11 (9):8892–8900.

Tang, Qing, Zhen Zhou, and Panwen Shen. 2012. Are MXenes Promising Anode Materials for Li Ion Batteries Computational Studies on Electronic Properties and Li Storage Capability of Ti_3C_2 and $Ti_3C_2X_2$ (X = F, OH) Monolayer. *Journal of the American Chemical Society* 134 (40):16909–16916.

Tao, Quanzheng, Martin Dahlqvist, Jun Lu, et al. 2017. Two-Dimensional $Mo_{1.33}C$ MXene with Divacancy Ordering Prepared from Parent 3D Laminate with In-Plane Chemical Ordering. *Nature Communications* 8 (1):14949.

Tran, Minh H., Timo Schäfer, Ali Shahraei, et al. 2018. Adding a New Member to the MXene Family: Synthesis, Structure, and Electrocatalytic Activity for the Hydrogen Evolution Reaction of V_4C_3Tx. *ACS Applied Energy Materials* 1 (8):3908–3914.

Turker, Furkan, Omer R. Caylan, Naveed Mehmood, Talip S. Kasirga, Cem Sevik, and Goknur Cambaz Buke. 2020. CVD Synthesis and Characterization of Thin Mo_2C Crystals. *Journal of the American Ceramic Society* 103 (10):5586–5593.

Urbankowski, Patrick, Babak Anasori, Taron Makaryan, et al. 2016. Synthesis of Two-Dimensional Titanium Nitride Ti_4N_3 (MXene). *Nanoscale* 8 (22):11385–11391.

VahidMohammadi, Armin, Ali Hadjikhani, Sina Shahbazmohamadi, and Majid Beidaghi. 2017. Two-Dimensional Vanadium Carbide (MXene) as a High-Capacity Cathode Material for Rechargeable Aluminum Batteries. *ACS Nano* 11 (11):11135–11144.

Wang, X., C. Garnero, G. Rochard, et al. 2017. A New Etching Environment (FeF_3/HCl) for the Synthesis of Two-Dimensional Titanium Carbide MXenes: A Route towards Selective Reactivity vs. Water. *Journal of Materials Chemistry A* 5 (41):22012–22023.

Wang, Lin, Liyong Yuan, Ke Chen, et al. 2016. Loading Actinides in Multilayered Structures for Nuclear Waste Treatment: The First Case Study of Uranium Capture with Vanadium Carbide MXene. *ACS Applied Materials & Interfaces* 8 (25):16396–16403.

Wang, Libo, Heng Zhang, Bo Wang, et al. 2016. Synthesis and Electrochemical Performance of Ti3C2Tx with Hydrothermal Process. *Electronic Materials Letters* 12 (5):702–710.

Wang, Ying, Wei Zheng, Peigen Zhang, Wubian Tian, Jian Chen, and ZhengMing Sun. 2019. Preparation of $(V_x,Ti_{1-x})_2C$ MXenes and Their Performance as Anode Materials for LIBs. *Journal of Materials Science* 54 (18):11991–11999.

Wu, Yuting, Ping Nie, Jiang Wang, Hui Dou, and Xiaogang Zhang. 2017. Few-Layer MXenes Delaminated via High-Energy Mechanical Milling for Enhanced Sodium-Ion Batteries Performance. *ACS Applied Materials & Interfaces* 9 (45):39610–39617.

Xie, Xiaohong, Yun Xue, Li Li, et al. 2014. Surface Al Leached Ti_3AlC_2 as a Substitute for Carbon for Use as a Catalyst Support in a Harsh Corrosive Electrochemical System. *Nanoscale* 6 (19):11035–11040.

Xu, Chuan, Libin Wang, Zhibo Liu, et al. 2015. Large-Area High-Quality 2D Ultrathin Mo_2C Superconducting Crystals. *Nature Materials* 14 (11):1135–1141.

Xuan, Jinnan, Zhiqiang Wang, Yuyan Chen, et al. 2016. Organic-Base-Driven Intercalation and Delamination for the Production of Functionalized Titanium Carbide Nanosheets with Superior Photothermal Therapeutic Performance. *Angewandte Chemie International Edition* 55 (47):14569–14574.

Yang, Jian, Michael Naguib, Michael Ghidiu, et al. 2016. Two-Dimensional Nb-Based M_4C_3 Solid Solutions (MXenes). *Journal of the American Ceramic Society* 99 (2):660–666.

Yazdanparast, Sanaz, Sina Soltanmohammad, Annika Fash-White, Garritt J. Tucker, and
    Geoff L. Brennecka. 2020. Synthesis and Surface Chemistry of 2D TiVC Solid-
    Solution MXenes. *ACS Applied Materials & Interfaces* 12 (17):20129–20137.
Zhang, Tian, Limei Pan, Huan Tang, et al. 2017. Synthesis of Two-Dimensional $Ti_3C_2T_x$
    MXene Using HCl+LiF Etchant: Enhanced Exfoliation and Delamination. *Journal of
    Alloys and Compounds* 695:818–826.
Zhao, Shuangshuang, Xing Meng, Kai Zhu, et al. 2017. Li-Ion Uptake and Increase in
    Interlayer Spacing of Nb_4C_3 MXene. *Energy Storage Materials* 8:42–48.
Zhao, Wei-Na, Na Yun, Zhen-Hua Dai, and Ye-Fei Li. 2020. A High-Performance Trace
    Level Acetone Sensor Using an Indispensable V_4C_3Tx MXene. *RSC Advances* 10
    (3):1261–1270.
Zhou, Jie, Xianhu Zha, Fan Y. Chen, et al. 2016. A Two-Dimensional Zirconium Carbide by
    Selective Etching of Al_3C_3 from Nanolaminated $Zr_3Al_3C_5$. *Angewandte Chemie
    International Edition* 55 (16):5008–5013.
Zhou, Jie, Xianhu Zha, Xiaobing Zhou, et al. 2017. Synthesis and Electrochemical Properties
    of Two-Dimensional Hafnium Carbide. *ACS Nano* 11 (4):3841–3850.
Zhu, Kai, Yuming Jin, Fei Du, et al. 2019. Synthesis of Ti_2CTx MXene as Electrode
    Materials for Symmetric Supercapacitor with Capable Volumetric Capacitance.
    Journal of Energy Chemistry 31:11–18.

3 Morphology and Microstructure of MXene

Nongfeng Huang and Zuzeng Qin
School of Chemistry and Chemical Engineering, Guangxi University

Hongbing Ji
Fine Chemical Industry Research Institute, School of Chemistry, Sun Yat-Sen University

CONTENTS

3.1 INTRODUCTION

MXenes are prepared by different methods, which have a significant effect on their morphology and microstructure. For example, multilayer MXene can be obtained using the HF as an etchant, while monolayer MXene can be obtained using LiF-HCl as an etchant. This chapter will discuss the morphology and microstructure of MXene prepared by different methods.

3.2 THE EFFECT OF FLUORIDE ETCHING ON THE MORPHOLOGY AND MICROSTRUCTURE OF MXENES

At present, most of the synthesis of MXenes is still obtained by HF etching the MAX phase. However, due to HF's toxicity, pollution, high risk, and operation difficulty, scientists are prompted to search for softer and safer etching agents. So far, some reduce or replace synthesis methods of HF have been studied, including combination of fluoride salt (LiF, NaF, KF or NH_4F) and HCl/H_2SO_4 to development in situ HF method (Lipatov et al. 2016), the use of transition metal fluoride, i.e., FeF_3 and CoF_2 (Wang et al. 2017; Cockreham et al. 2019), using the fluoride reagent, i.e., base compound (Xuan et al. 2016; Li et al. 2018; Sheng et al. 2018), and other methods, i.e., the electrochemical etching (Sun et al. 2017; Sheng et al. 2018), and molten-salt etching (Li, Shao, et al. 2020). The etchant showed a significant influence on the surface chemistry, oxidation sensitivity, hydrophilicity, lamination ability, electrical conductivity, the nature of cations inserted, slice size, and degree of the defects (Sun et al. 2017; Sheng et al. 2018; Li et al. 2020).

Although many reports have used HF etching for MXenes, different etching conditions, such as HF concentration, reaction temperature, and duration, vary greatly; variations are mainly caused by the differences in the initial structure and the physical and chemical properties of the MXenes powders, such as the particle size, crystallinity, and purity. Therefore, it is difficult to compare the Mxenes from different reports directly, and we need understand the effects of different etching methods on the MXenes morphology and the physical and chemical properties. We also need to know how to control the properties of MXenes by simply adjusting the synthetic conditions.

Experimental studies have shown that using high HF concentration (Ti_3C_2-HF48) would result in multiple layers of MXenes separated by large gaps (acscordion-like structure), while using low HF concentration (Ti_3C_2-HF10) would result in tightly stacked slices (Alhabeb et al. 2017). As the reaction of Al and HF produces H_2 in the etching process, the reaction becomes faster when HF concentration increases, thus contributing to the expansion of the interlayer structure. Therefore, using high concentration HF in the stripping process would lead to more macroscopic defects and form multilayer MXenes with good separation (Alhabeb et al. 2017). For the LiF etching samples, no matter their synthetic conditions, they are more easily stratified, and the suspension can be shaken (Benchakar et al. 2020).

However, no methods have been found to precisely control the surface functional groups, which are highly dependent on synthetic pathways and synthetic postprocessing. [1]H and ^{19}F nuclear magnetic-resonance (NMR) studies have successfully quantified the $Ti_3C_2T_x$ MXenes single bonds prepared via the HF and HCl-LiF routes. Ti_3C_2 was prepared by using the volatile HF route and the milder HCl-LiF route. There were some significant differences between the two preparation methods: SEM images show that although HF etched samples exhibited an accordion-like morphology, the samples etched by the LiF and HCl mixture showed a more compact structure with no significant stratification. XRD patterns show that the (0002) plane of the sample by LiF-HCl etching shifted to a lower angle (corresponding to the c parameter of approximately 25); when compared to the HF etching MXene, the space distance is 19–20 Å. However, there is evidence of a

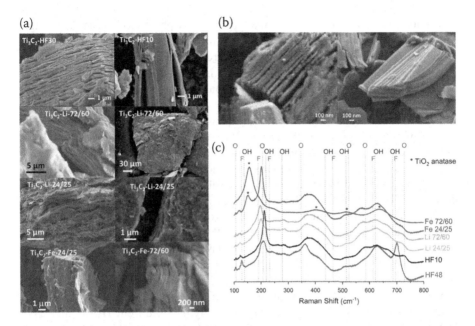

FIGURE 3.1 (a) Typical SEM micrographs of the different synthesized MXenes; SEM micrograph of, (b) Ti_3C_2-Fe-72/60 sample at high magnification showing, (c) Raman spectra of the as-synthesized MXenes (Benchakar et al. 2020).

small amount of unreacted Ti_3AlC_2 in the LiF-HCl etched samples (Hope et al. 2016). Mohamed Benchakar analyzed the influence of different etching agents on the surface chemistries of $Ti_3C_2T_x$ multilayers obtained from the same MAX phase batch. Studies have shown that Ti_3C_2-Fe-72/60 (etched by FeF_3 and HCl for 72 h at 60 °C) is highly oxidized (confirmed by the Raman spectrum (Figure 3.1c)) because of the high temperature. Furthermore, the SEM images of the sample (Figure 3.1b) also confirmed the existence of a large number of surface nanoparticles, which may be TiO_2 anatase. However, no such nanoparticles were observed on the nonoxidized samples, such as Ti_3C_2-HF10 (10 wt% for Ti_3C_2-HF10, in Figure 3.1b) (Benchakar et al. 2020). The presence of these nanoparticles significantly increased the specific surface area of Ti_3C_2-FE-72/60 from 2–15 $m^2 \cdot g^{-1}$ (other samples) to 75 $m^2 \cdot g^{-1}$ (Benchakar et al. 2020). Due to the oxidation of MXene, the O fraction of FeF_3 etched samples was significantly higher than that of other samples and correspondingly increased to the fraction of $TiO_{2-x}F_{2x}$ (Figure 3.1c); the C fraction was also increased, in particular with the Ti_3C_2-Fe-72/60 sample, confirming the formation of amorphous carbon phases during the etching process.

3.3 THE EFFECT OF ALKALI-ASSISTED ETCHING ON THE MORPHOLOGY AND MICROSTRUCTURE OF MXENES

Most MXene synthesis methods inevitably involve HF. In the MAX phase, the metal M-A bond is usually weaker than the M-X bond; therefore, in most situations, selectively etching layer A is the first choice (Sun et al. 2004). To rapidly remove A (elements A, such as Al and Si) from a typical MAX phase, such as Ti_3AlC_2, high concentration of fluoride ions (for example, in HF aqueous solution of 50 wt%) must be used as an etching agent due to the strong binding capacity of F. Although convenient and effective, these processes are harmful to the environment and degrade material properties due to inert F terminals, so new fluorine-free methods for removing layer A are urgently needed.

From the thermodynamic point of view, phase MAX and Ti_3AlC_2 can react with OH reliably. For example, Li (Li et al. 2017) successfully prepared $Ti_3C_2(OH)_2$ by etching Ti_3AlC_2 precursor with KOH in the presence of a small amount of water, which was caused by the formation of some oxide/hydroxide layers on the surface of Ti_3AlC_2 due to the kinetic limitations. XRD results show a certain amount of oxide on the initial Ti_3AlC_2 powder, such as Al_2O_3 (Xuan et al. 2016). However, the yield of $Ti_3C_2T_x$ obtained by continuous sodium-hydroxide (NaOH) treatment was still low, even after removing the oxide cover with an HF solution of 10 wt%, which suggests that the formation of new Al(oxide) hydroxyl layers with alkali exposure, such as Al $(OH)_3$ and AlO(OH), may further hinder the required Al extraction. Therefore, these protective layers must be cleaned. Inspired by Bayer process, Yang (Li et al. 2018) prepared $Ti_3C_2T_x$ (T=OH, O) by etching Ti_3AlC_2 with NaOH-assisted hydrothermal method. The results show that Al can be selectively removed from Ti_3AlC_2 in a 27.5 M NaOH solution at 270 °C, and the multilayer $Ti_3C_2T_x$ terminated by OH and O with a purity of 92 wt% can be obtained. SEM analysis shows that the $Ti_3C_2T_x$ (270 °C, 27.5 M NaOH) is a dense, layered structure, which is different from the common accordion-like $Ti_3C_2T_x$ obtained by a high concentration (50 wt%) HF treatment. On the contrary, it is very similar in $Ti_3C_2T_x$ using the low concentration (5 wt%) HF with a slow H_2 formation rate. In this structure, the specific surface area of $Ti_3C_2T_x$ (about 16 m^2/g) is slightly larger than that of the original Ti_3AlC_2 (about 9 m^2/g), but smaller than that of $Ti_3C_2T_x$ prepared with the 50 wt% HF (about 23 m^2/g). Energy dispersive X-ray spectroscopy (EDS) results show that Ti, C, O, and Na are uniformly distributed on the flaky surface with a small amount of al appearance, which means that most of the Al layer is eliminated. In addition, TEM (Figure 3.2a) shows that these $Ti_3C_2T_x$ films had an uneven interval of about 1.2 nm between the layers, which is also greater than the original Ti_3AlC_2 (about 0.93 nm), and similar to HF-$Ti_3C_2T_x$ with Na insertion. These results are also supported by the high-angle circular dark-field scanning transmission electron microscope (HAADF-STEM), as shown in Figure 3.2b.

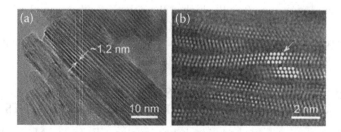

FIGURE 3.2 Microstructure characterizations of the $Ti_3C_2T_x$ flakes via, (a) TEM, and (b) HAADF-STEM, respectively. The imposed bright spots in Tipositions (Li et al. 2018).

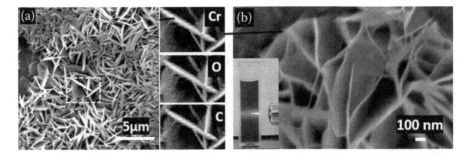

FIGURE 3.3 (a) Low-magnification and (b) high-magnification SEM images with the elemental mapping analysis of Cr_2CT_x (Inset: digital photo of Cr_2CT_x well-dispersed in water demonstrating a clear Tyndall scattering effect) (Pang et al. 2019).

3.4 EFFECTS OF ELECTROCHEMICAL ETCHING ON THE MORPHOLOGY AND MICROSTRUCTURE OF MXENES

There are other ways to avoid using HF for etching, and the most promising one may be electrochemical synthesis (Verger et al. 2019). MXene (Ti_2CT_x) prepared by dilute HCl electrochemical etching is a safer and milder method than the traditional HF etching method. Its structure and surface properties will be changed with the change of electrochemical etching conditions. The parent phase is treated in an HCl, or ammonium chloride, or tetramethylammonium hydroxide (TMAOH) electrolyte to remove the Al. Yang (Yang et al. 2018) obtained more than 90% of $Ti_3C_2T_x$ single-layer and double-layer high yields from Ti_3AlC_2 using binary water electrolytes, a process that has also been successfully applied to produce other Mxenes, such as V_2CT_x and Cr_2CT_x. In addition, it has effectively solved the long-standing problem of high concentration of HF and showed its potential as a general preparation method for MXene. On the other hand, MXene synthesized by the nonelectrochemical etching method can reach 25 μm and has a flower-like structure (Pang et al. 2019), showed in Figure 3.3. Other methods, including hydrothermal synthesis to produce $Ti_3C_2T_x$ (Li et al. 2018), and synthetic first nitride $Ti_4N_3T_x$ in molten salt of LiF, NaF, and KF (Urbankowski et al. 2016), and fully Cl-terminated $Ti_3C_2Cl_2$ and Ti_2CCl_2 MXene were synthesized in $ZnCl_2$ Lewis acidic molten salt (Yu et al. 2019), are also used for synthesizing the MXene.

3.5 EFFECTS OF DIFFERENT MXENE MORPHOLOGIES ON PHOTOCATALYTIC REACTION

In the etching process, the reaction between Al and HF produces H_2, and when the HF concentration increases, the reaction is faster and expands the interlamellae structure. However, it is very difficult to meet the needs of researchers that layer spacing only produced in the etching process. For the LiF etching samples, simply shaking the suspension can realize MXenes stratified; otherwise, adopting high-polymer intercalating layered material, such as dimethyl sulfoxide, would be used for separation of the layers. Consider these facts: (i) MXene has many hydrophilic functional groups, i.e., -OH and -O, on its surface, and it is easy for MXene to establish strong connections with various semiconductors; (ii) MXene shows excellent metal conductivity, which ensures effective carrier transfer; (iii) exposure of terminal metal sites on MXene, such as Ti, Nb, or V, may lead to greater redox reactivity than carbon materials; (iv) the presence of a large number of hydrophilic functional groups on MXene promotes its strong interaction with water molecules; (v) MXene can play a stable role in aqueous solution and other outstanding properties; therefore, MXene is expected to become a promising material in photocatalysis field (Ran et al. 2017).

At present, MXene has been widely used as a cocatalyst to enhance the photocatalytic activities of TiO_2, g-C_3N_4, CdS, ZnS, and $Zn_xCd_{1-x}S$ (Su et al. 2019a). In the Ti_3C_2/CdS system, a Schottky junction is formed between the Ti_3C_2 and the CdS, which can be used as an electron trap to capture photogenerated electrons effectively. Under illumination, electrons are excited from the valence band (VB) to the conduction band (CB) of CdS, and then photogenerated electrons migrate from CB of cadmium sulfide to Ti_3C_2 due to Ti_3C_2 which has a low Fermi level. Furthermore, Ti_3C_2 has excellent electrical conductivity and suitable H adsorption Gibbs free energy, and the photogenerated electrons concentrate on Ti_3C_2 would effectively reduce protons to form hydrogen (Ran et al. 2017).

MXenes first appeared with a layered structure, and then modified stripping was carried out to meet the researcher's needs; most of the multilayer or nanoparticle particles of MXene were first applied in the field of photocatalysis, which experienced slow charge transfer and showed good photocatalytic activity. For example, when multilayer 5 wt% $Ti_3C_2T_x$ is used as the cocatalyst of TiO_2, the hydrogen production, 17.8 $\mu mol \cdot h^{-1} \cdot g_{cat}^{-1}$, of TiO_2/$Ti_3C_2T_x$ composites in photocatalytic hydrogen evolution reaction is 400% higher than that of pure rutile TiO_2 (Wang et al. 2016). When $Ti_3C_2T_x$ and Cu are used as cocatalysts at the same time to improve the photocatalytic activity of TiO_2, Cu/TiO_2@$Ti_3C_2T_x$ efficient photocatalyst split water to produce hydrogen at 860 $\mu mol \cdot h^{-1} \cdot g_{cat}^{-1}$, which significantly improves the yield (Peng et al. 2018).

3.5.1 MULTILAYERED MXENES

In the multilayer Ti_3C_2-based photocatalyst, Ti_3C_2 is used as a promising platform for assembling the semiconductors and electron receptors to facilitate photogenerated charge transfer and separation (Thang Phan et al. 2020), the introduction of multilayer Ti_3C_2 can determine a positive role in a range of photocatalytic applications. For example, Li (Li, Zhao, et al. 2020) prepared g-C_3N_4/Ti_3C_2 by one-step calcination of melamine as a precursor, and the 2D layered g-C_3N_4 is uniformly distributed on the multilayer Ti_3C_2 surface, forming a special 2D/3D structure (Figure 3.4a). Due to the excellent conductivity of Ti_3C_2, the intimate interface between g-C_3N_4 and Ti_3C_2, and the construction of a Schottky barrier to capture photogenerated electrons, the heterojunction shows unique activity in photocatalytic H_2 formation (Figure 3.4b). Yang (Yang et al. 2020) used urea as a precursor and intercalator to mix with multilayer Ti_3C_2 and calcined at 550 °C in N_2, and this process obtained ultrathin 2D/2D Ti_3C_2/g-C_3N_4 heterojunction in which the

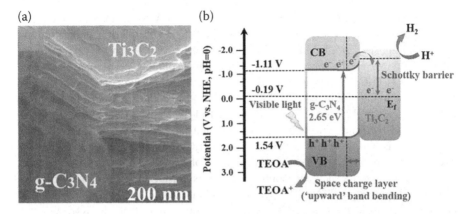

FIGURE 3.4 (a) SEM images, (b) schematic diagram of the proposed mechanism for photocatalytic H_2 production over g-C_3N_4/Ti_3C_2 composite (Li, Zhao, et al. 2020).

micromolecule urea not only acted as a gas template to peel multilayer Ti_3C_2 into nanometer but also acted as the precursor of melamine to form g-C_3N_4. This method can prevent the tedious process of Ti_3C_2 stripping and greatly increase the production of ultra-thin Ti_3C_2. The CO_2 photoreduction activity test showed that pure g-C_3N_4 (UCN) showed very weak photoactivity; however, when Ti_3C_2 is combined with g-C_3N_4, the photocatalytic activity is greatly improved. The optimum samples (10TC) showed the yields of CO and CH_4 were 5.19 and 0.044 mol·h^{-1}·g^{-1}, respectively, and the total CO_2 conversion rate was 8.1 times that of UCN alone (Li, Zhao, et al. 2020).

TiO_2 semiconductor has become one of the most commonly used materials due to its chemical stability, cost effectiveness, light resistance, and strong redox capability (Xiang, Yu, and Jaroniec 2012). However, some inherent defects of TiO_2 have hindered its application, including low visible light response with a bandgap of 3.0–3.2 eV, rapid photoexcited electron-hole pair recombination, and slow kinetics of surface H_2 release reaction (Cai et al. 2019). Considered a promising cocatalyst for photocatalytic production of H_2 instead of Pt (Guo et al. 2016), Ti_3C_2@TiO_2 composites fabricated by adding additional exogenous titanium were reported in the photocatalysis aspect (Luo et al. 2018; Ye et al. 2018). It is worth noting that Ti_3C_2 MXenes can be used as a powerful carrier and as a precursor for the in situ growth of TiO_2 photocatalyst. Peng (Peng et al. 2016) synthesized an atomic scale interfacial heterojunction comprising the (001) exposed TiO_2 nanosheets inserted in the layered Ti_3C_2 via a partial hydrothermal oxidation of Ti_3C_2 assisted by $NaBF_4$.

Multilayer Ti_3C_2 is the titanium source formed by TiO_2, which forms a Schottky junction with the surface of TiO_2 (001), and the surface junction between surfaces (001) and (101) on the TiO_2 nanosheet and the tight interface Schottky junction between TiO_2 and Ti_3C_2 facilitate the spatial separation of photogenerated electrons and holes, leading to significantly enhanced photocatalytic degradation ability of methyl orange. Based on the above advantages, (001) TiO_2/Ti_3C_2 hybrid was further modified to construct a visible photocatalytic hydrogen evolution ternary composite. Li (Li, Ding, Yin, et al. 2020) grew TiO_2 nanosheets on conductive

Ti$_3$C$_2$ MXenes in situ, and uniformly distributed 1T-WS$_2$ nanoparticles on the TiO$_2$@Ti$_3$C$_2$ composite, obtaining a unique 1T-WS$_2$@TiO$_2$@Ti$_3$C$_2$ bimetallic co-catalyst composite, and the 1T phase content reached 73%. When the optimal WS$_2$ ratio of 15 wt%, the photocatalytic hydrogen evolution performance of 1T-WS$_2$@TiO$_2$@Ti$_3$C$_2$ composites is nearly 50 times than that of TiO$_2$ nanosheets, which is due to the conductivity of Ti$_3$C$_2$ MXene and 1T-WS$_2$, improving the electron transfer efficiency (Li, Ding, Yin, et al. 2020). TiO$_2$ nanosheets (NSs) were grown in situ on Ti$_3$C$_2$ MXene with high conductivity, and then MoS$_2$ rich in molybdenum vacancies was uniformly distributed on TiO$_2$@Ti$_3$C$_2$ composites using the two-step hydrothermal method, and the unique structure of Mo$_x$S@TiO$_2$@Ti$_3$C$_2$ composite with molybdenum vacancy and double cocatalysts (Ti$_3$C$_2$ and MoS$_2$) was obtained (Li, Ding, Liang, et al. 2020). The hydrogen yield produced by Mo$_x$S@TiO$_2$@Ti$_3$C$_2$ composite photocatalysis is nearly 193 and six times than that of TiO$_2$ NSs and MoS$_2$@TiO$_2$@Ti$_3$C$_2$ composite. In Mo$_x$S@TiO$_2$@Ti$_3$C$_2$ composites, the introduction of molybdenum vacancies into MoS$_2$ can produce a large number of active sites with an enhanced specific activity. The existence of molybdenum vacancy inhibits the recombination of carriers and is beneficial to the reaction.

In summary, the above study confirms the emerging role of multilayer Ti$_3$C$_2$ as an essential platform for integration with other semiconductor nanomaterials, resulting in solid interface contact, excellent light collection, improved electrical conductivity, and highly efficient hybrid structures based on Ti$_3$C$_2$. Advantageous charge transfer and separation in various photocatalytic applications results. However, the co-catalytic activity and stability of Ti$_3$C$_2$ remain inadequate (for example, the precipitation activity of H$_2$ is lower than that of Pt), and the surface engineering design of multilayer Ti$_3$C$_2$ is critical but challenging.

3.5.2 SINGLE LAYER MXENES

Compared with block structure, two-dimensional layered materials can create a large number of photocatalytic reaction sites due to their larger specific surface area, and the two-dimensional feature can extend the life of photogenerated electrons and holes by reducing the required electron migration distance before reaching the solid. Therefore, the synthesis of single-layer or low-layer MXene with larger surface area and more interfaces is of great significance for obtaining effective Ti$_3$C$_2$-based photocatalysts.

Monolayer cocatalysts have a larger specific surface area and more exposed active sites, shortening the migration distance of photogenerated charge. The high conductivity and ultrathin characteristics of single-layer Ti$_3$C$_2$Tx are conducive to the migration of photogenerated electrons from TiO$_2$ to Ti$_3$C$_2$T$_x$, realizing higher electron-hole separation efficiency and higher photocatalytic hydrogen generation than the multilayer Ti$_3$C$_2$T$_x$ as a cocatalyst. The photocatalytic hydrogen production rate (2.65 mmol·h^{-1}·g$_{cat}^{-1}$) of the optimized single-layer Ti$_3$C$_2$T$_x$/TiO$_2$ composite material was more than 9 times higher than that of pure TiO$_2$ and 2.5 times higher than that of multilayer TiO$_2$. The significantly enhanced activity was attributed to the excellent conductivity of single-layer Ti$_3$C$_2$T$_x$ and carrier separation at the MXene/TiO$_2$ interface (Su et al. 2019b).

For 2D semiconductors, there are two assembly forms, including vertical growth and face-to-face, to construct 2D/2DTi$_3$C$_2$-based photocatalysts. Zhao (Zuo et al. 2020) reported that a single layer Ti$_3$C$_2$ MXenes nanometer sheet (MNs) synthesized by HF/HCl was used to support the vertical ultrathin 2D ZnIn$_2$S$_4$ nanometer sheet (UZNs) to form a sandwich-layered heterogeneous structure (UZNs-MNS-UZNs), as shown in Figure 3.5a,b. UZNs at the appropriate lateral epitaxy on the surface of MN simultaneously expand the specific surface area, improve pore size and hydrophilicity, and thus promote the release of photocatalytic H$_2$. The H$_2$ release rate of MNZIS-4 composite is 6.6 times higher than that of the original ZnIn$_2$S$_4$, as shown in Figure 3.5c. The electron-density distribution and work function between ZnIn$_2$S$_4$ and Ti$_3$C$_2$ are calculated, and the formation of the Schottky junction is confirmed. The ultrathin two-dimensional structure and the presence of Schottky barrier can effectively inhibit the electron-hole recombination under photoexcitation, and promote the charge transfer and separation under photoexcitation, as shown in Figure 3.5d. In addition, the firm "face to face" interaction of multilayer or single-layer Ti$_3$C$_2$ MXenes and the effective coordination of

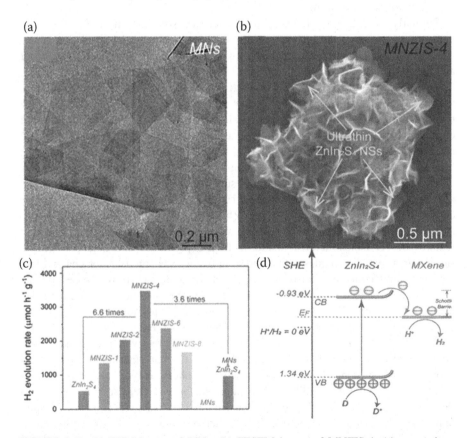

FIGURE 3.5 (a) TEM image of MNs, (b) FESEM image of MNZIS-4, (c) comparison of PHE activity for ZnIn$_2$S$_4$, MNZIS-X, MNs, and physical mixture MNs and ZnIn$_2$S$_4$, (d) illustration of the band structure of MNZIS-4 (Zuo et al. 2020).

obvious interface exposure can improve photocatalytic activity and positively affect
the photocatalytic activity stabilizing the photocatalytic performance (Yang et al.
2019). Cao (Cao et al. 2018) used the ultrasonic stripping method to etch Ti_3AlC_2
and prepared a two-dimensional ultrathin Ti_3C_2 nanometer tablet of MXenes. A
novel 2D/2D heterojunction Ti_3C_2/Bi_2WO_6 ultra-thin nanosheet was successfully
prepared by in situ growing Bi2WO6 ultrathin nanosheet on the surface Ti_3C_2 ultrathin nanosheet (Figure 3.6). The results show that the ultrathin Ti_3C_2/Bi_2WO_6
nanosheet 2D/2D heterojunction significantly improves the photocatalytic performance of CO_2 reduction under simulated solar irradiation. The total yield of CH_4
and CH_3OH of the optimized Ti_3C_2/Bi_2WO_6 hybrid material is 4.6 times than that
of the original Bi_2WO_6 ultrathin nanosheets, which, caused by the Ti_3C_2/Bi_2WO_6
2D/2D heterojunction, possesses large interface contact area and quite short charge-
transport distance, leading to the efficient electron transfer from the photocatalyst
(Bi_2WO_6) to cocatalyst (Ti_3C_2), as shown in Figure 3.6j. Moreover, the improved
specific surface area and pore structure of the 2D/2D heterostructured ultrathin
nanosheets assure a distinctly enhanced CO_2 adsorption capability, which further
boosts the photocatalytic reactions.

Similarly, 2D/2D g-C_3N_4/Ti_3C_2 with a unique structure showed similar advantages in hydrogen production reaction, CO_2 reduction, and degradation of organic pollutants. The interface contact area between the layered Ti_3C_2 nanosheet
and the 2D semiconductor is maximized to form a well-structured Schottky junction, conducive to the separation of photogenerated electrons and holes. In addition,
the active site of stratified Ti_3C_2 was fully exposed at the same time; therefore, the
photocatalytic activity of two-dimensional/two-dimensional heterostructure based
on layered Ti_3C_2 can be improved (Ran et al. 2017; Su et al. 2019a). For 1D/2D
heterogeneous structures, the contact between the main photocatalyst and the
layered Ti_3C_2 is line-surface contact. Although the interface area of heterogeneous
junctions is smaller than that of 2D/2D, such hybrid dimensional heterojunctions
have attracted a lot of research attention. 1D photocatalysts are mainly due to their
unique intrinsic characteristics (Hou and Zhang 2020). 1D nanorods and nanowires
catalysts are thought to have greater advantages in bulk materials. The mentioned
features of 1D nanostructures include a high aspect ratio and a larger specific
surface area, which provides a fast and long distance charge-transfer channel and
enhances light collection performance (Ge et al. 2016). Therefore, based on the
ultrathin heterogeneous junction of 2D MXene, the method of combining 1D
photocatalyst with 2D Ti_3C_2 nanometer sheet is expected to obtain controllable
photocatalytic performance with high efficiency (Li, Li, et al. 2020).

A CdS-Ti_3C_2 MXenes (CM) binary heterostructure, which combines one-
dimensional CdS nanorods onto a single-layer Ti_3C_2, has clear nanostructure and
strong interface contact and can effectively photocatalyze H_2 precipitation
(Figure 3.7a) (Xiao et al. 2020). The photocatalytic H_2 release rate of CM-20
sample was 7 times higher than that of the original CdS nanorods (Figure 3.7b) due
to its high conductivity, unique two-dimensional morphology, full-spectrum optical
capture of the hydrophilic Ti_3C_2 nanorods, and close interface contact. The band
diagram of 1D CdS was determined by the Mott-Schottky diagram and UPS analysis (Figure 3.7c). In the case that the Fermi level of Ti_3C_2 is higher than the CdS

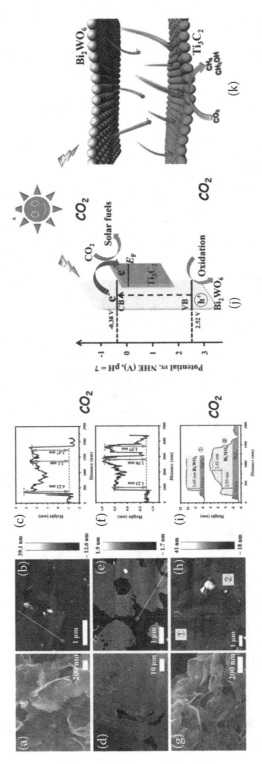

FIGURE 3.6 (a–c) Typical FESEM, AFM images, and height cutaway view of Bi_2WO_6, (d–f) Ti_3C_2 nanosheets and (g–i) TB2. (j) Energy-level structure diagram of Bi_2WO_6 and Ti_3C_2; (k) Photo-induced electron transfer process at the interface of the hybrids (Cao et al. 2018).

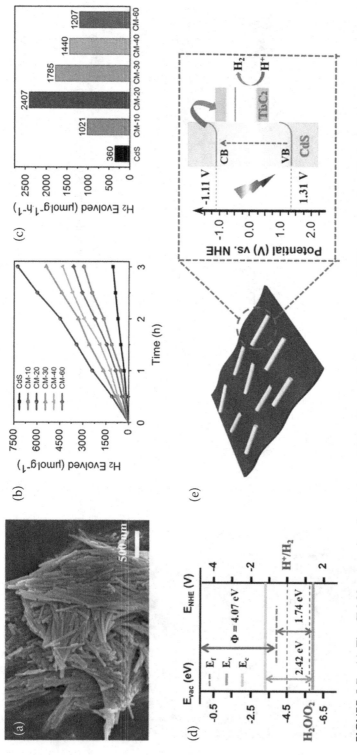

FIGURE 3.7 (a) The CM-20 composite, (b, c) photocatalytic H_2 evolution perfromance of different samples, (d) the band energy diagram and (e) representative schematic of 1D CdS/2D MXene Schottky heterojunction for hydrogen production (Xiao et al. 2020).

conduction band, the Schottky barrier is constructed to accelerate charge separation and inhibit the return of photogenerated electrons to the CdS conduction band (Figure 3.7d). The Schottky-based photocatalyst suggests a synergistic effect between one-dimensional n-type CdS and two-dimensional Ti_3C_2 nanosheet. In addition, CdS/Ti_3C_2 heterostructure photocatalysts have solid interactions and can inhibit photocorrosion. For example, due to the strong interaction between CdS and Ti_3C_2, Cd^{2+} ions released by CdS photo etching are restricted in situ in the Ti_3C_2 layer, preventing THE leaching of Cd^{2+} without damaging the surface reaction sites, thus enabling CdS to be photocorrosion-resistant (Xie et al. 2018).

All in all, single or multilayers of Ti_3C_2 as an emerging expandable body to support hydrothermal and electrostatic self-assembly low-dimensional semiconductor, is a kind of get-efficient photocatalyst of Ti_3C_2 advantageous method; the main reason is that a layered Ti_3C_2 nano piece can provide more interface contact and active site, and is suitable for surface engineering design. However, due to the instability of single or multilayer Ti_3C_2 exposed to water and oxygen for a long time, high performance and catalytic durability of Ti_3C_2 nanocomposites need to be designed.

3.5.3 MXENE QUANTUM DOTS

More recently, the emergence of MXene quantum dots (MQD) has brought more attention to photocatalysis, sensors, and energy conversion. Compared with GQD, MQD is expected to have good water solubility and more flexible application. In addition, The preparation of MQD is simple, easy to expand, and has a high yield. As one of the most commonly used MXenes materials, Ti_3C_2 MXenes have excellent photocatalytic performance and can be used in environmental governance and solar power generation. Therefore, Ti_3C_2 QDs are a competitive candidate material in photocatalysis.

Compared with 2D Ti_3C_2 MXene nanocrystals, 0D Ti_3C_2 MXene quantum dots have a larger specific surface area and more prosperous active edge sites. Thus, Ti_3C_2 MXene QD enhances light collection and acts as an electron acceptor to improve photogenic carrier transfer in photocatalysts, i.e., Ti_3C_2 QDs were successfully synthesized by the solvent, thermal method (PEI) and were an effective cocatalyst combined with $g-C_3N_4$ for photocatalytic production of H_2 (Li et al. 2019). The optimized $g-C_3N_4@Ti_3C_2$ QD composite (5.5 wt%) achieved a significant increase in visible photocatalytic H_2 production (5111.8 $mol \cdot g^{-1} \cdot h^{-1}$, which was 26, 3, and 10 times higher than that of pure $g-C_3N_4$ nanocrystals, $Pt/g-C_3N_4$, and Ti_3C_2 MXenes/g-C_3N_4, respectively. The high-efficiency photocatalytic activity of $g-C_3N_4@Ti_3C_2$ quantum dot hybridization is mainly attributed to the microparticle size of the Ti_3C_2 quantum dot promoter, which can increase specific surface area and provide more active sites. In addition, the firm contact between metal Ti_3C_2 quantum dots and $g-C_3N_4$ nanometers facilitates the transfer of photogenerated electrons from $g-C_3N_4$ to Ti_3C_2 quantum dots. Thus, electrons can rapidly participate in the photocatalytic H_2 release. Zeng (Zeng et al. 2019) formed a layered heterostructure of Ti_3C_2 QD decorated Cu_2O nanowire (NWs) (Figure 3.8a), which was assembled on copper mesh by simple electrostatic self-assembly to reduce CO_2 to methanol by photocatalysis. In the practical photocatalytic reduction of CO_2, the methanol production rate of Ti_3C_2

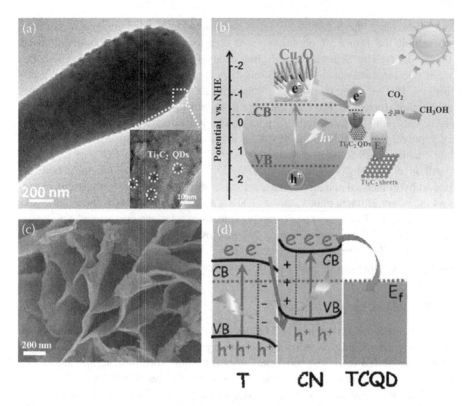

FIGURE 3.8 (a) TEM image of Ti_3C_2 QDs/Cu_2O NWs/Cu heterostructure with a magnified region shown in inset. (b) Energy-level diagram of Ti_3C_2 QDs/Cu_2O NWs/Cu and Ti_3C_2 sheets/Cu_2O NWs/Cu heterostructures (Zeng et al. 2019). (c) SEM image of TiO_2/C_3N_4/Ti_3C_2 QDs (T-CN-TC) and (d) transfer of charges in S-scheme heterojunction of T-CN-TC (He et al. 2020).

QDs/Cu_2O NWS/Cu (153.38 ppm/cm^2) is 8.25 and 2.15 times that of Cu_2O NWS/Cu and Ti_3C_2 Sheets/Cu_2O NWs/Cu, respectively. Experiments and density functional-theory calculations show that the addition of Ti_3C_2 quantum dots can significantly improve the stability and photocatalytic performance of Cu_2O NWs by enhancing charge transfer, carrier density, photo adsorption, and decreasing band bending edge and charge recombination (Figure 3.8b).

He (He et al. 2020) reported a 2D/2D/0D TiO_2/C_3N_4/Ti_3C_2 QD (TCQD) hetero-junction flower-like photocatalyst for CO_2 reduction (Figure 3.8c), using self-assembly technology such as telomere TCQD synthesized in C_3N_4 surface modification of TiO_2/C_3N_4 core-shell structure composite materials (2D/2D/0D TiO_2/ C_3N_4/ Ti_3C_2 QD). This mechanism obtains zero distance contact between ultrathin core-shell nano-crystals and strong coupling between C_3N_4 and TCQD, providing an effective transport channel for carriers. It is worth noting that the S-shaped heterojunction be-tween TiO_2 and C_3N_4 helps maintain the redox capacity of photogenic carriers (Figure 3.8d). Furthermore, the presence of TCQD accelerates the spatial migration of electrons on C_3N_4 CB and becomes the "highway" and acceptor for electron transport. Therefore, the S-Scheme heterojunction and TCQD were highly synergistic to enhance the photocatalytic carbon dioxide reduction site. The CO generation rate was 4.39 $\mu mol \cdot g^{-1} \cdot h^{-1}$, which was more than three times of the CO production rate of C_3N_4 or titanium dioxide alone. This work provides broad prospects for potential applications of Ti_3C_2 quantum dots as photocatalytic auxiliary catalysts (e.g., quantum constraints, edge effects, adjustable surface functions, and low Fermi levels).

Compared with 2D Ti_3C_2 MXene sheets, 0D Ti_3C_2 MXene QD has attracted more attention due to its better water solubility and more prosperous active edge sites. In addition, like 2D Ti_3C_2 MXene tablets, Ti_3C_2 QD can also be used as an electron acceptor to increase carrier transport. The 0D quantum dots (QDS) derived from 2D materials corresponding to 2D materials (e.g., graphene QDS, Boron nitrates nanorods, and molybdenum disulfide nanorods) have unique properties such as wider bandgaps due to quantum constraints, better tunability of physical and chemical properties, more prosperous active edge sites, and better dispersion. Therefore, Ti_3C_2 QD can be used as a catalyst in electrical/photocatalysis.

In conclusion, the morphology and microstructure of MXenes prepared by different methods have significant effects. HF etchers are generally used to prepare multilayer MXene, while LiF/HCl etchers are used to prepare single-layer MXene. Some new methods, such as base-compound assisted etching and electrochemical etching, are used to avoid the harmful effects of inert F ions on the environment. However, the yields obtained by these methods are relatively low, so the high-yield, safe, and nontoxic methods are still worth exploring. MXene materials used to light catalytic reaction showed interesting catalytic activity; compared with the multilayer, single layer has a larger specific surface area and a large number of photocatalytic reaction sites. However, single or multilayer Ti_3C_2 is unstable when exposed to water and oxygen for a long time, so nanocomposites with high performance and catalytic durability of Ti_3C_2 should be designed. Compared with 2D Ti_3C_2 MXene nanosheets, 0D Ti_3C_2 MXene quantum dots have wider band gap, better physicochemical properties, more abundant active edge sites, and better dispersion, which make it become one of the promising photocatalytic materials.

ACKNOWLEDGMENTS

This work was supported by the National Natural Science Foundation of China (22078074, 21938001), Guangxi Natural Science Foundation (2019GXNSFAA-245006, 2020GXNSFDA297007, 2016GXNSFFA380015), Special funding for 'Guangxi Bagui Scholars', and Scientific Research Foundation for High-level Personnel from Guangxi University.

REFERENCES

Alhabeb, Mohamed, Kathleen Maleski, Babak Anasori, et al. 2017. Guidelines for Synthesis and Processing of 2D Titanium Carbide ($Ti_3C_2T_x$ MXene). *Chemistry of Materials* 29 (18):7633–7644.

Benchakar, Mohamed, Lola Loupias, Cyril Garnero, et al. 2020. One MAX Phase, Different MXenes: A Guideline to Understand the Crucial Role of Etching Conditions on $Ti_3C_2T_x$ Surface Chemistry. *Applied Surface Science* 530:147209.

Cai, Jingsheng, Jiali Shen, Xinnan Zhang, et al. 2019. Light-Driven Sustainable Hydrogen Production Utilizing TiO_2 Nanostructures: A Review. *Small Methods* 3 (1):1800184.

Cao, Shaowen, Baojia Shen, Tong Tong, Junwei Fu, and Jiaguo Yu. 2018. 2D/2D Heterojunction of Ultrathin MXene/Bi_2WO_6 Nanosheets for Improved Photocatalytic CO_2 Reduction. *Advanced Functional Materials* 28 (21):1800136.

Cockreham, Cody B., Xianghui Zhang, Houqian Li, Ellis Hammond-Pereira, and Di Wu. 2019. Inhibition of AlF_3 Center Dot 3H(2)O Impurity Formation in $Ti_3C_2T_x$ MXene Synthesis under a Unique CoF_x/HCl Etching Environment. *ACS Applied Energy Materials* 1 (11):8145–8152.

Ge, Mingzheng, Chunyan Cao, Jianying Huang, et al. 2016. A Review of One-Dimensional TiO_2 Nanostructured Materials for Environmental and Energy Applications. *Journal of Materials Chemistry A* 4 (18):6772–6801.

Guo, Zhonglu, Jian Zhou, Linggang Zhu, and Zhimei Sun. 2016. MXene: A Promising Photocatalyst for Water Splitting. *Journal of Materials Chemistry A* 4 (29):11446–11452.

He, Fei, Bicheng Zhu, Bei Cheng, Jiaguo Yu, Wingkei Ho, and Wojciech Macyk. 2020. 2D/2D/0D TiO_2/C_3N_4/Ti_3C_2 MXene Composite S-Scheme Photocatalyst with Enhanced CO_2 Reduction Activity. *Applied Catalysis B-Environmental* 272:119006.

Hope, Michael A., Alexander C. Forse, Kent J. Griffith, et al. 2016. NMR Reveals the Surface Functionalisation of Ti_3C_2 MXene. *Physical Chemistry Chemical Physics* 18 (7):5099–5102.

Hou, Huilin and Xiwang Zhang. 2020. Rational Design of 1D/2D Heterostructured Photocatalyst for Energy and Environmental Applications. *Chemical Engineering Journal* 395:125030.

Li, Yujie, Lei Ding, Yichen Guo, Zhangqian Liang, Hongzhi Cui, and Jian Tian. 2019. Boosting the Photocatalytic Ability of g-C_3N_4 for Hydrogen Production by Ti_3C_2 MXene Quantum Dots. *Acs Applied Materials & Interfaces* 11 (44):41440–41447.

Li, Yujie, Lei Ding, Zhangqian Liang, Yanjun Xue, Hongzhi Cui, and Jian Tian. 2020. Synergetic Effect of Defects Rich MoS_2 and Ti_3C_2 MXene as Cocatalysts for Enhanced Photocatalytic H_2 Production Activity of TiO_2. *Chemical Engineering Journal* 383:123178.

Li, Yujie, Lei Ding, Shujun Yin, et al. 2020. Photocatalytic H_2 Evolution on TiO_2 Assembled with Ti_3C_2 MXene and Metallic 1T-WS_2 as Co-Catalysts. *Nano-Micro Letters* (1):63–74.

Li, Jing-Yu, Yue-Hua Li, Fan Zhang, Zi-Rong Tang, and Yi-Jun Xu. 2020. Visible-Light-Driven Integrated Organic Synthesis and Hydrogen Evolution over 1D/2D CdS-$Ti_3C_2T_x$ MXene Composites. *Applied Catalysis B-Environmental* 269:118783.

Li, Youbing, Hui Shao, Zifeng Lin, Jun Lu, and Qing Huang. 2020. A General Lewis Acidic Etching Route for Preparing MXenes with Enhanced Electrochemical Performance in Non-Aqueous Electrolyte. *Nature Materials* 19 (8):894–899.

Li, Gengnan, Li Tan, Yumeng Zhang, Binghan Wu, and Liang Li. 2017. Highly Efficiently Delaminated Single-Layered MXene Nanosheets with Large Lateral Size. *Langmuir* 33 (36):9000–9006.

Li, Tengfei, Lulu Yao, Qinglei Liu, et al. 2018. Fluorine-Free Synthesis of High Purity $Ti_3C_2T_x$ (T = OH, O) via Alkali Treatment. *Angewandte Chemie International Edition* 57 (21):6115–6119.

Li, Jinmao, Li Zhao, Shimin Wang, Jin Li, Guohong Wang, and Juan Wang. 2020. In Situ Fabrication of 2D/3D g-C_3N_4/Ti_3C_2 (MXene) Heterojunction for Efficient Visible-Light Photocatalytic Hydrogen Evolution. *Applied Surface Science* 515:145922.

Lipatov, A., M. Alhabeb, M. R. Lukatskaya, A. Boson, Y. Gogotsi, and A. Sinitskii. 2016. Effect of Synthesis on Quality, Electronic Properties and Environmental Stability of Individual Monolayer Ti_3C_2 MXene Flakes. *Advanced Electronic Materials* 2 (12):1600255.

Luo, Qiang, Bo Chai, Mengqiu Xu, and Qizhou Cai. 2018. Preparation and Photocatalytic Activity of TiO_2-Loaded Ti_3C_2 with Small Interlayer Spacing. *Applied Physics a-Materials Science & Processing* 124 (7):495.

Pang, Sin-Yi, Yuen-Ting Wong, Shuoguo Yuan, et al. 2019. Universal Strategy for HF-Free Facile and Rapid Synthesis of Two-Dimensional MXenes as Multifunctional Energy Materials. *Journal of the American Chemical Society* 141 (24):9610–9616.

Peng, Chao, Ping Wei, Xiaoyao Li, Yunpeng Liu, and Kangle Lv. 2018. High Efficiency Photocatalytic Hydrogen Production over Ternary $Cu/TiO_2@Ti_3C_2T_x$ Enabled by Low-Work-Function 2D Titanium Carbide. *Nano Energy* 53:97–107.

Peng, Chao, Xianfeng Yang, Yuhang Li, Hao Yu, Hongjuan Wang, and Feng Peng. 2016. Hybrids of Two-Dimensional Ti_3C_2 and TiO_2 Exposing {001} Facets toward Enhanced Photocatalytic Activity. *Acs Applied Materials & Interfaces* 8 (9):6051–6060.

Ran, Jingrun, Guoping Gao, Fa-Tang Li, Tian-Yi Ma, Aijun Du, and Shi-Zhang Qiao. 2017. Ti_3C_2 MXene Co-Catalyst on Metal Sulfide Photo-Absorbers for Enhanced Visible-Light Photocatalytic Hydrogen Production. *Nature Communications* 8:13907.

Sheng, Yang, Panpan, et al. 2018. Fluoride-Free Synthesis of Two-Dimensional Titanium Carbide (MXene) Using A Binary Aqueous System. *Angewandte Chemie* 57 (47): 15491–15495.

Su, Tongming, Zachary D. Hood, Michael Naguib, et al. 2019a. 2D/2D Heterojunction of Ti_3C_2/g-C_3N_4 Nanosheets for Enhanced Photocatalytic Hydrogen Evolution. *Nanoscale* 11 (17):8138–8149.

Su, Tongming, Zachary D. Hood, Michael Naguib, et al. 2019b. Monolayer $Ti_3C_2T_x$ as an Effective Co-Catalyst for Enhanced Photocatalytic Hydrogen Production over TiO_2. *ACS Applied Energy Materials* 2 (7):4640–4651.

Sun, Z., D. Music, R. Ahuja, S. Li, and J. M. Schneider. 2004. Bonding and Classification of Nanolayered Ternary Carbides. *Physical Review B – Condensed Matter and Materials Physics* 70 (9):092102

Sun, Wanmei, Smit Shah, Yexiao Chen, et al. 2017. Electrochemical Etching of Ti_2AlC to Ti_2CT_x (MXene) in Low-Concentration Hydrochloric Acid Solution. *Journal of Materials Chemistry A* 5:21663–21668.

Thang Phan, Nguyen, Nguyen Dinh Minh Tuan, Tran Dai Lam, et al. 2020. MXenes: Applications in Electrocatalytic, Photocatalytic Hydrogen Evolution Reaction and CO_2 Reduction. *Molecular Catalysis* 486:110850.

Urbankowski, Patrick, Babak Anasori, Taron Makaryan, et al. 2016. Synthesis of Two-Dimensional Titanium Nitride Ti_4N_3 (MXene). *Nanoscale* 8 (22):11385–11391.

Verger, Louisiane, Varun Natu, Michael Carey, and Michel W. Barsoum. 2019. MXenes: An Introduction of Their Synthesis, Select Properties, and Applications. *Trends in Chemistry* 1 (7):656–669.

Wang, Xiaoyao, Cyril Garnero, Guillaume Rochard, et al. 2017. A New Etching Environment (FeF_3/HCl) for the Synthesis of Two-Dimensional Titanium Carbide MXenes: A Route towards Selective Reactivity vs Water. *Journal of Materials Chemistry A* 5 (41): 22012–22023.

Wang, Hui, Rui Peng, Zachary D. Hood, Michael Naguib, Shiba P. Adhikari, and Zili Wu. 2016. Titania Composites with 2D Transition Metal Carbides as Photocatalysts for Hydrogen Production under Visible-Light Irradiation. *Chemsuschem* 9 (12):1490–1497.

Xiang, Quanjun, Jiaguo Yu, and Mietek Jaroniec. 2012. Synergetic Effect of MoS_2 and Graphene as Cocatalysts for Enhanced Photocatalytic H_2 Production Activity of TiO_2 Nanoparticles. *Journal of the American Chemical Society* 134 (15):6575–6578.

Xiao, Rong, Chengxiao Zhao, Zhaoyong Zou, et al. 2020. In Situ Fabrication of 1D CdS Nanorod/2D Ti_3C_2 MXene Nanosheet Schottky Heterojunction toward Enhanced Photocatalytic Hydrogen Evolution. *Applied Catalysis B-Environmental* 268:118382.

Xie, Xiuqiang, Nan Zhang, Zi-Rong Tang, Masakazu Anpo, and Yi-Jun Xu. 2018. $Ti_3C_2T_x$ MXene as a Janus Cocatalyst for Concurrent Promoted Photoactivity and Inhibited Photocorrosion. *Applied Catalysis B-Environmental* 237:43–49.

Xuan, Jinnan, Zhiqiang Wang, Yuyan Chen, et al. 2016. Organic-Base-Driven Intercalation and Delamination for the Production of Functionalized Titanium Carbide Nanosheets with Superior Photothermal Therapeutic Performance. *Angewandte Chemie* 55 (47):14569–14574.

Yang, C., Q. Y. Tan, Q. Li, et al. 2020. 2D/2D Ti_3C_2 MXene/g-C_3N_4 Nanosheets Heterojunction for High Efficient CO_2 Reduction Photocatalyst: Dual Effects of Urea. *Applied Catalysis B-Environmental* 268:11.

Yang, Yang, Zhuotong Zeng, Guangming Zeng, et al. 2019. Ti_3C_2 Mxene/Porous g-C_3N_4 Interfacial Schottky Junction for Boosting Spatial Charge Separation in Photocatalytic H_2O_2 Production. *Applied Catalysis B-Environmental* 258:117956.

Yang, Sheng, Panpan Zhang, Faxing Wang, et al. 2018. Fluoride-Free Synthesis of Two-Dimensional Titanium Carbide (MXene) Using A Binary Aqueous System. *Angewandte Chemie-International Edition* 57 (47):15491–15495.

Ye, Minheng, Xin Wang, Enzuo Liu, Jinhua Ye, and Defa Wang. 2018. Boosting the Photocatalytic Activity of P25 for Carbon Dioxide Reduction by Using a Surface-Alkalinized Titanium Carbide MXene as Cocatalyst. *Chemsuschem* 11 (10):1606–1611.

Yu, Lianghao, Zhaodi Fan, Yuanlong Shao, Zhengnan Tian, Jingyu Sun, and Zhongfan Liu. 2019. Versatile N-Doped MXene Ink for Printed Electrochemical Energy Storage Application. *Advanced Energy Materials* 9 (34):1901839.

Zeng, Zhiping, Yibo Yan, Jie Chen, Ping Zan, Qinghua Tian, and Peng Chen. 2019. Boosting the Photocatalytic Ability of Cu_2O Nanowires for CO_2 Conversion by MXene Quantum Dots. *Advanced Functional Materials* 29 (2):1806500.

Zuo, Gancheng, Yuting Wang, Wei Liang Teo, et al. 2020. Ultrathin $ZnIn_2S_4$ Nanosheets Anchored on $Ti_3C_2T_x$ MXene for Photocatalytic H_2 Evolution. *Angewandte Chemie-International Edition* 59 (28):11287–11292.

4 Property of MXene

Nongfeng Huang, Xuan Luo, and Zuzeng Qin
School of Chemistry and Chemical Engineering, Guangxi University

Hongbing Ji
Fine Chemical Industry Research Institute, School of Chemistry, Sun Yat-Sen University

CONTENTS

4.1 INTRODUCTION

The functional groups (such as –O, –OH, –F) play an important role in the property of MXene. For example, the electrical conductivity, work function, and surface properties of the MXenes can be regulated by altering the surface-functional groups. In this chapter, the influence factor on the property of MXene will be summarized systematically.

4.2 THE EFFECT OF FUNCTIONAL GROUPS ON THE PROPERTY OF MXENE

MXenes is a general term for two-dimensional carbides or nitride materials, which usually have graphene. The general chemical formula is $M_{n+1}X_nT_x$, where M is the transition metal element (Ti, Zr, V, Cr, Sc, etc), X is C or N, T_x is the surface group (O^{2-}, OH^-, F^-, NH_3^-, NH_4^+, etc.), and n is generally 1 to 3. MXenes is made of

DOI: 10.1201/9781003156963-4

layered ternary metallic carbides/nitrides (MAX phase) that are chemically ex-
tracted from elements "A" (groups IIIA and IVA, such as aluminum). Element A
(usually Al) is usually etched with A fluorine-containing acid solution, such as HF,
NH_4HF_2, or lithium fluorine-hydrochloric acid. This results in A mixture of oxygen-
containing and fluorine-containing surface terminals, whose reaction equation is:

$$M_{n+1}AlX_n + 3HF = AlF_3 + 3/2H_2 + M_{n+1}X_n \tag{4.1}$$

$$M_{n+1}X_n + 2H_2O = M_{n+1}X_n(OH)_2 + H_2 \tag{4.2}$$

$$M_{n+1}X_n + 2HF = M_{n+1}X_nF_2 + H_2 \tag{4.3}$$

Reaction (4.1) is the primary and most basic reaction, and then reaction (4.2) and
reaction (4.3) occur. From (4.1), the reaction generates AlF_3, Al, stripped out,
MXenes to separation, reaction (4.2) and (4.3) adhered on the surface of the MXenes
OH and F group, and the generation of the surface groups and surface metal Ti
MXenes excess electrons; a direct combination with surface greatly decreases, thus
forming a stable nanometer sheet form. Therefore, the attached OH and F groups are
the most basic units of MXenes (Naguib et al. 2011). The surface of MXenes is
randomly distributed by OH, F, and O functional groups, which is difficult to control
experimentally (Caffrey 2018). Moreover, the coverage area of its functional groups
and the distribution rule of its functional groups are not clear, but it has been confirmed experimentally that OH and O coexist on the surface of MXenes (Naguib et al.
2011). Li et al. (Li et al. 2018) argued that MXene was different from other two-
dimensional materials in that its functional groups did not depend on any controlled
reaction. However, they were generated by the stripping process itself, indicating that
the functional terminating group is related to the synthesis method. For example, the
functional groups of samples synthesized by HF and LiF-HCl are significantly different. HF etching mainly produces functional groups containing –F, while HCl and
LiF treatment mainly produces materials containing oxygen-containing surface groups
(Hope et al. 2016).

Due to the high electronegativity of O or F atoms on the end group, the functionalized MXenes have a negative surface charge. Therefore, through strong
Coulomb interaction between MXene and cations, MXene can be reversible electrochemical intercalation with cations such as Na^+, Mg^{2+}, and Al^{3+} (Lukatskaya
et al. 2013; Kajiyama et al. 2016). In the H_2SO_4 electrolyte, the surface groups also
play an important role in the redox reaction, in which the oxygen groups can be
converted to hydroxyl, and the protons and electrons can be regulated at the same
time (Hu et al. 2016). This example indicates that surface groups have an important
influence on the redox reaction, so it is necessary to adjust the proportion of different groups in different reactions to meet their needs. For example, $Mo_2TiC_2T_x$
and $Mo_2Ti_2C_3T_x$ MXenes can be treated with tetrabutylammonium hydroxide
(TBAOH) to reduce fluorine content and increase oxygen content (Anasori et al.
2016). The cationic organic molecules can be easily inserted in between the negative charge on the surface of multilayer MXene, allowing an effective layered 2D

chip, as well as in the alkaline-solution alkaline MXene made changes F groups in the $Ti_3C_2T_x$ OH groups (Luo et al. 2016), or by hydrazine-embedded two-dimensional titanium carbide (Ti_3C_2 MXene); MXenes moisture content decreased significantly, and the hydroxyl and fluoride amount is significantly reduced, at the end to change its surface chemical (Mashtalir et al. 2016).

4.2.1 THE EFFECT OF FUNCTIONAL GROUPS ON THE BAND STRUCTURE AND CONDUCTIVITY OF MXENES

The surface-functional groups of MXene affect its electrochemical properties and affect its electronic properties (including energy-band structure, work function) and optical properties. Although no methods have been found to precisely control the types of surface-functional groups, those surface-functional groups are highly dependent on synthetic pathways and synthetic post-processing. ^1H and ^{19}F nuclear magnetic-resonance (NMR) studies have successfully quantified $Ti_3C_2T_x$ MXenes single bonds prepared via HF and HCl-LiF routes (Hope et al. 2016). Figure 4.1a and b show the $Ti_3C_2T_x$ MXene single bond prepared by the volatile HF route and the mild HCl-LiF route, respectively, in different forms of multilayer $Ti_3C_2T_x$. The

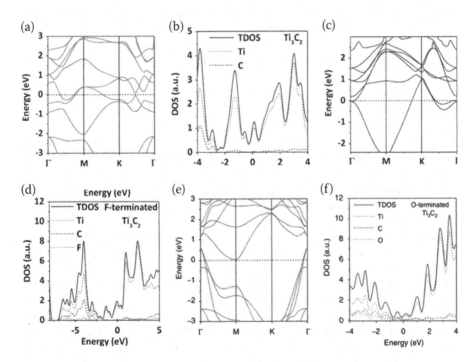

FIGURE 4.1   (a) The electronic band structure and (b) the total density of states (TDOS) and partial density of states (PDOS) for pure Ti_3C_2, (c) the electronic band structure, and (d) the TDOS and PDOS for F-terminated Ti_3C_2, (e) the calculated band structure of O-terminated Ti_3C_2, (f) the total density of states (TDOS) and partial density of states (PDOS) for O-terminated Ti_3C_2 (Ran et al. 2017).

energy band structures of Ti_3C_2, Ti_3C_2 at F end, and Ti_3C_2 at O end were determined by density functional theory. Figure 4.1c and d shows that pure Ti_3C_2 exhibits metallic properties with many electronic states across the Fermi level. In contrast, Ti_3C_2 of the F group (Figure 4.1c and d) and Ti_3C_2 of the O group (Figure 4.1e and f) show a reduced number of states at the Fermi level, indicating a lower conductivity. However, the continuous electronic states of F-terminated Ti_3C_2 and O-terminated Ti_3C_2 across the Fermi level indicate that their conductivity is still good. Thus, Ti_3C_2, which contains many functional groups, exhibits excellent electrical conductivity, which means it has an extraordinary ability to transmit electrons. This unique advantage of MXene makes it an excellent co-catalyst with better performance than similar products such as graphene and carbon nanotubes, whose conductivity decreases significantly after termination with –O, –OH, and –COO. In addition, the Fermi levels of Ti_3C_2, O, and –F functional groups are 0.05, 1.88, and 0.15 V, respectively. Among them, Ti_3C_2 at –O terminal shows the maximum positive value of E_F, which means that it has the strongest ability to accept photoelectrons from semiconductor photocatalysts.

Many theoretical studies have been carried out to identify the electron-band structure and properties of MXene from its large family (Khazaei et al. 2017). It is expected that most of the functional MXene have metal/semimetal band structures, while a few MXene systems are semiconductors. According to reports, the band gaps of Sc_2CT_2 (T = O, OH, F), Ti_2CO_2, Zr_2CO_2, and Hf_2CO_2 are between 0.24 and 1.8 eV (Khazaei et al. 2013). The reported band structure and band gaps are shown in Figure 4.2. Only $Sc_2C(OH)_2$ is expected to have a direct band gap, while the others are predicted to have indirect band gaps among those six MXenes. Interestingly, the biaxial strains of Ti_2CO_2, Zr_2CO_2, and Hf_2CO_2 are 4%, 10%, and 14%, respectively, which can cause indirect to direct band-gap transitions (Yu et al. 2015). The transition from bare MXene (metal) to functionalized MXene is accompanied by the opening of band gaps due to new states or O $2p$ orbitals resulting from strong band hybridization between M 3D orbitals and C $2p$ below the Fermi level (Magne et al. 2015), explaining by the lower electronegativity of transition metals compared to functional groups and carbon atoms. The OH and F groups can receive one electron, while the O group can receive two electrons from the transition metal atom.

In addition, Ti, Zr, and Hf are in the same column of the periodic table with the same shell electronic structure (two electrons in the s and d orbitals), so a similar character transition trend from metal to a semiconductor is found in metals, and the corresponding M_2C MXene is formed after the O-group functionalization. The band gap (M = Ti, Zr, Hf) of M_2CO_2 increases with the increase of the number of metal (M) atoms due to the decrease of metal electronegativity (Gandi, Alshareef, and Schwingenschloegl 2016). For $Ti_{n+1}C_nO_2$ (n ≥ 2) MXene, the contribution of O $2p$ orbitals decreases with the increase of n, and such MXene is no longer a semiconductor (Xie and Kent 2013). It is reported that Mo_2C and Cr_2C are semiconductors with OH, F, and Cl functional groups, while several dual-transition metals MXene are O-terminated semiconductors. Liu et al. (Liu, Xiao, and Goddard 2016) proposed MXenes as the Schottky barrier-free contact material for semiconductor TMD. For example, The O-terminal MXene is suitable for the Schottky

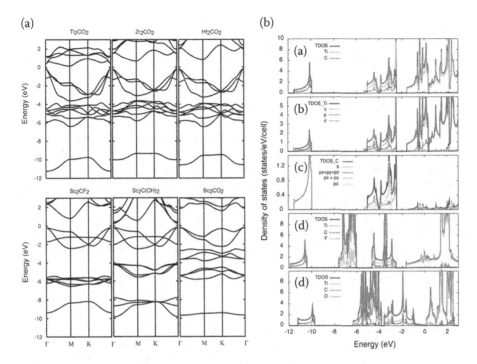

FIGURE 4.2 (a) Band structure of semiconducting MXene systems, (b) DOS of Ti$_2$C, (c, d) the PDOS on atomic orbitals of Ti and C atoms, respectively. (e, f) DOSs of Ti$_2$CF$_2$, and Ti$_2$CO$_2$, respectively, Fermi energy is at zero. For the semiconductors, it is shifted to the center of the gap (Khazaei et al. 2013).

barrier-free hole injection of P-type WSe$_2$. The proposed concept is based not only on the high-power function of O-terminal MXene, but also on the weak van der Waals interaction at the metal-semiconductor interface, which can release the Fermi level nailing effect (Liu, Stradins, and Wei 2016). Similarly, MXene with a low-power function on OH sealing end is suitable for N-type MoS$_2$ contact.

In metal-semiconductor junctions, the Mott-Schottky rule is usually used to predict or interpret contact conditions. Depending on the difference between the work function (metal) and the electron-affinity or ionization energy (semiconductor), an ohmic contact or Schottky junction with a certain barrier height can be achieved. However, the experimental results often deviate from the basic theory, mainly due to interface defects and the Fermi level napping effect. For example, electron-beam evaporation in high-energy metal deposition methods damages the underlying material and creates a gap state. Recently, it has been shown that the transferred metal electrode can follow the Mott-Schottky rule (Liu et al. 2018) to achieve the Van Der Waal metal-semiconductor junction. In operation, the MoS$_2$ transistor with a variety of transfer metal electrodes shows an entirely different operating behavior compared to the case of using ordinary high-energy deposition metal contacts. Therefore, low-energy metal deposition methods are important, and solid solution MXenes may be one of the practical options.

DFT calculations show that the band gap of MXene can be modified by changing the surface ends. For example, unterminated Ti_3C_2 MXene essentially acts as a metal conductor, while OH or F-terminated MXenes are semiconductors with band gaps of 0.05 and 0.1 eV, respectively (Naguib et al. 2011). Thus, with changing the band gap (Xu et al. 2016; Zha et al. 2016), these materials can be used for various applications, from field-effect transistors to semiconductors. The band gap of MXene can also change with the change of element composition of the m layer. Anasori et al. (Anasori et al. 2016) demonstrated that the topmost two layers of Ti in $Ti_3C_2(OH)_2$ were replaced by Mo layers, and a two-dimensional double-layer TMC of $Mo_2TiC_2(OH)_2$ was produced, in which Mo atoms gave MXene semiconductor performance with a narrow band gap of 0.05 eV (Anasori et al. 2016) and the transition is confirmed by the positive and negative magnetoresistance values of $Ti_3C_2(OH)_2$ and $Mo_2TiC_2(OH)_2$ at 10 K (Anasori et al. 2016; Anasori, Luhatskaya, and Gogotsi 2017).

To further understand the structure of MXene, it is very important to analyze the position of surface groups. Establishing precise structures can help achieve this goal (Hu et al. 2015). It is predicted that the possible location of the surface group is between three adjacent C atoms above the hollow site (Tang, Zhou, and Shen 2012). This model proved too simple, as subsequent studies reported that the location and orientation of terminals should be more complex (Khazaei et al. 2013), with proper surface functionalization, Sc_2C, Ti_2C, Zr_2C, and Hf_2C MXene expected to become semiconductors. Theoretically, it can be concluded that functional Cr_2C and Cr_2N MXene are magnetic. Thermoelectric calculation based on Boltzmann theory shows that semiconductor MXene can obtain a very large Seebeck coefficient at a low temperature. The ultimate properties of surface-terminating groups are influenced by their types and the constituent elements of MXene.

In most cases, the structure of MXenes is reduced to a unified terminating group, which does not satisfy the actual situation. Therefore, more accurate models are needed to represent this complex system since different surface groups can coexist and, in many cases, random adsorption can occur together. Furthermore, the reported existence of multiple layers of stacking, in reality, reveals the importance of the analysis of interlayer interactions (including hydrogen bonds and van der Waals forces) (Khazaei et al. 2013). In addition, the surface groups of MXenes not only affect their stability but also affect almost all other properties, including electronic and optical properties (Kim, Wang, and Alshareef 2019). Theoretically, the energy-band structure, state density, and other electronic properties have been studied deeply. To date, first-principles calculations and experimental results have revealed that MXenes should be metallic or semimetallic, whereas semiconductor MXenes represent only a small number. Early theoretical studies reported several examples, as shown in Figure 4.2, including Sc_2CT_2 (T = OH, F, O), Ti_2CO_2, Zr_2CO_2, and Hf_2CO_2, with band gaps between 0.24 and 1.8 eV (Khazaei et al. 2013). In particular, band gaps of $Sc_2C(OH)_2$, Sc_2CF_2, and Sc_2CO_2 were reported to be 0.45, 1.03, and 1.8 eV, respectively, reflecting the influence of surface groups on the electronic properties of MXenes. In addition, compared with Sc_2CO_2, the energy band structures of $Sc_2C(OH)_2$ and Sc_2CF_2 are more similar. That is because both the F and OH groups need to receive one more electron to be stable, and O needs to receive two more electrons to be stable. Through the analysis and prediction of its band structure, it is found that five of the six structures have indirect band gaps, with the exception of $Sc_2C(OH)_2$.

To further understand the mechanism of semiconduction after functionalization of some MXenes, the electronic structures of pure MXenes and functionalized MXenes were studied (Khazaei et al. 2013). The results show that, similar to the MAX phase, the exposed MXene is metallic because the Fermi energy level is all contributed by the d orbital of the M element (Figure 4.2a) (Khazaei et al. 2013, 2017). The p band of element X lies below the d band, and the band gap is very small. Specifically, the functionalization of F causes Fermi energy to move down because each F takes an electron from the system. Instead, the O functionalization leads to further downward movement, which ultimately makes Ti_2CO_2 a semiconductor. The projection DOS(PDOS) of Ti_2C, Ti_2CF_2, and Ti_2CO_2 in Figure 4.2 can make this mechanism clearer. Further studies have also reported that strain (Lee, Cho, and Chung 2014; Yu et al. 2015), or applied electric field (Lee et al. 2014; Li 2016) strongly affect the band gaps of Ti_2CO_2 and Sc_2CO_2.

In addition, the similarity of M elements will reveal some similar electronic properties. Typical examples are Ti, Zr, and Hf. Since they are in the same group, their outermost shell has the same number of electrons and configuration ($4s^2$, $5s^2$, and $6s^2$, respectively), allowing them to show a similar tendency to shift from metal to semiconductor properties (Kim, Wang, and Alshareef 2019). For M_2CO_2 (M = Ti, Zr, Hf) MXenes, with the decrease of M metallicity, the forbidden bandwidth increases continuously (Gandi, Alshareef, and Schwingenschloegl 2016). For $Ti_3C_2O_2$ and $Ti_4C_3O_2$ MXenes, the increase of Ti (M) element number cancels out the effect of O's $2p$ orbital, so these MXenes revert to metal again (Xie and Kent 2013).

As mentioned earlier, in most cases, MXenes with two different transition metals have a randomly distributed solid solution structure. There are some special cases where these two M elements are ordered. These structures can be represented in the general form M'2 M" C_2T_n and M'2 M" $_2C_3T_n$, since M' only replaces the surface layer and M" forms the central layer (Anasori et al. 2015). A typical example is $Mo_2TiC_2T_n$ because Mo is on the surface and Ti is in the center. Compared with random distribution or monolayer structure, the ordered dual-transition metal structure (ordered dual M) has different electronic properties. For example, $Mo_2TiC_2T_n$, which is similar in structure to Ti_3C_2Tn, is semiconductive rather than metallic (Seh et al. 2016).

Further studies have speculated that some ordered double M MXenes, $M'_2M"C_2O_2$ and $M'_2M"_2C_3O_2$ (where M' = Mo, W and M" = Ti, Zr, Hf) with O ends are 2D topological insulators and topological semimetals, respectively (Mohammad Khazaei et al. 2016). In addition, ordered double MMXene and single MMXene can be topological insulators, such as W_2CO_2, Mo_2CO_2, and Cr_2CO_2 (Weng et al. 2015). In theory, some $Mo_2M"C_2O_2$ (M" = Ti, Zr, Hf) MXene can even produce spin Hall effect at room temperature because they have a quite large topological gap (0.1–0.2 eV) (Si et al. 2016). Although the synthesis of nontrivial topological MXenes is still difficult, these MXenes have high application potential, especially in superconductivity. Figure 4.3 shows the projected energy-band structure of $Sc_2C(OH)_2$, $Hf_2C(OH)_2$, Hf_2CF_2, Hf_2CO_2, graphene, BN, graphene, and MoS_2 (Mohammad Khazaei et al. 2016). The color and size of the points indicate the projected weight of the wave function on the different components. The

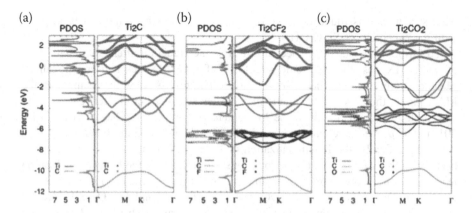

FIGURE 4.3 Projected density of states and projected band structures for (a) Ti$_2$C, (b) Ti$_2$CF$_2$, and (c) Ti$_2$CO$_2$ the Fermi energy is located at zero energy (Khazaei 2017).

Fermi level is set at 0 eV. As will be explained later, these states are mainly in the vacuum region outside the hydrogen atom. They have near-free electron (NFE) characteristics and have parabolic energy dispersion related to the crystal wave vector. The lowest energy of the NFE state of hydroxy-terminated MXenes is not occupied, located above and near the Fermi level, lower than that of graphene, BN, graphene, and molybdenum disulfide. The reason is that there are positively charged hydrogen atoms on the surface of OH-terminated MXenes (Mohammad Khazaei et al. 2016). Using such charged hydrogen atoms on the surface, hydroxy-terminated MXenes may have potential application value in the purification of heavy metals such as Pb and Hg, which indicates that hydroxy-terminated MXenes is easier to obtain than other 2D materials (Mohammad Khazaei et al. 2016).

Compared with OH terminal MXenes, F terminal and O terminal MXenes also show the changing trend of NFE state at high energy. In the figure, Hf$_2$CF$_2$, Hf$_2$CO$_2$ has the lowest NFE state at 3 eV and 5 eV above Fermi energy. This high energy is a major limitation of F and O terminal MXenes applications (Mohammad Khazaei et al. 2016). The partially occupied NFE state near the Fermi level is very sensitive to environmental changes. The state is reduced when great pressure is applied, gas molecules are adsorbed, or MXenes are doped with graphene, boron nitride, or graphene (Mohammad Khazaei et al. 2016).

4.2.2 THE EFFECT OF FUNCTIONAL GROUPS ON THE OXIDATION STABILITY OF MXENES

The stability of MXenes has a great impact on the reaction. For example, Lipatov et al. found that the conductivity decreased exponentially after the edge oxidation of MXenes (Lipatov et al. 2016). Therefore, understanding the longevity of MXenes is particularly important for its future development in various fields.

Zhang (Zhang, Pinilla, et al. 2017) studied the factors that may affect the stability of Ti$_3$C$_2$T$_x$ colloids and the schemes to improve their stability, showing the results showed in SEM images and corresponding histograms (Figure 4.4a,b). After one

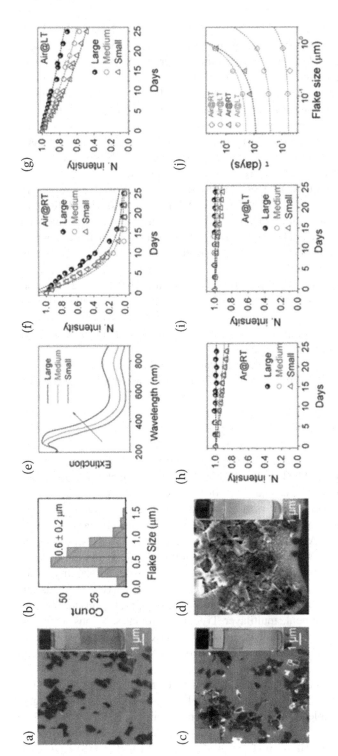

FIGURE 4.4 (a) Standard d-Ti$_3$C$_2$T$_x$ MXene solution SEM image and (b) histogram of flake size. The standard d-Ti$_3$C$_2$T$_x$ solution after aging in Air@RT for (c) 7 days and (d) 30 days. Insets in (a–c) show that the color of the standard solution changes over time oxidation stability of MXenes. (e) Normalized extinction spectra of fresh d-Ti$_3$C$_2$T$_x$ solutions with different flake sizes. (f–j) Stability results of the small, medium, and large d-Ti$_3$C$_2$T$_x$ flakes in Air@RT, Air@LT, Ar@RT, and Ar@LT, respectively.

week of aging in Air@RT, some "branches" formed at the edges, and 2–3 nm nanoparticles appeared on the substrate of the sheet. It later proved to be anatase. The SEM image in Figure 4.4c shows the contrast difference between the edge and the base surface position, and the thin slice at the edge is brighter. The colloid solution was further aged in Air@RT for 30 days, and the material was completely decomposed into anatase fragments and disordered carbon. During the aging process, the colloid solution also changed from green to dark white (Figure 4.4a, c, d). Electron microscopic analysis showed that the edge was more fragile than the base. Its aging is the process of moving from the edge to the center. Within two weeks, almost all carbide is oxidized to anatase-type titanium dioxide.

Then, the effect of particle size on the reaction was explored, as shown in Figure 4.4f–j (Zhang, Pinilla, et al. 2017). The stability of colloidal solutions of various sizes indicated that large MXene slices were the most stable, followed by medium and small slices. The degradation rate dependence on the nanosheet size also indicates the edge-driven reaction. Introducing Ar into LT can greatly improve the life of various colloidal solutions. For example, after 25 days, large MXene tablets degraded only 1.2% in Ar@LT, 1.4% in Ar@RT, and 29.1% and 98.5% in Air@LT and Air@RT, respectively. The reduced degradation rate in Ar@LT indicates that dissolved oxygen is the main oxidant, while the aqueous medium is a mild oxidant. Compared with large sheet samples, small sheet samples have a shorter lifetime in all environments.

To ensure the performance of MXene in various applications and improve its stability, Zhang et al (Zhang, Pinilla, et al. 2017) loaded an aqueous solution into a bottle filled with Ar, and the pressure inside the air-tight bottle was up to 2.3 mbar. After aging in AIR@RT for 8 h, the concentration of d-Ti_2CT decreased by 50% but completely disappeared after 24 h. Degradation of d-Ti_2CT slices follows a single exponential decay well. Obtained 0.4 days (10 h) time constant (τ) from the fitting line, which was reasonable because the solution aged 12 h in Air@RT changed to opaque white. TEM images show that many branches/nanoparticles have formed and are confirmed as anatase. After 12 h of storage in Ar@LT, the d-Ti_2CT slices were relatively clean with clear edges and shapes. Stability is significantly improved. The time constant (τ) in Ar@LT increased to 10 days, 24 times of the time constant in the Air@RT sample. Raman and XPS spectra also show that the life of colloidal Ti_2CT is greatly improved by storing d-Ti_2CT solution in Ar@LT, indicating that the proposed improved-method MXene solution stability is effective and universal. The reason is that compressed Ar effectively prevents air from entering the water medium and forming dissolved oxygen. In addition, before and during ultrasound treatment, multilayer $Ti_3C_2T_x$ colloidal solutions were degassed to remove dissolved oxygen from fresh d-$Ti_3C_2T_x$ aqueous solutions. Therefore, the solution stored in a bottle full of Ar is well protected from the effects of dissolved oxygen.

In addition, $Ti_3C_2T_x$ samples were prepared with different etching agents, investigating their stability and analyzing the morphological and composition changes of fresh samples and samples aged in room temperature air for 30 days (called 30-day samples) (Yang et al. 2020). Figure 4.5a–c shows transmission electron microscope (TEM) images of HF-$Ti_3C_2T_x$ at 30 days, LH- $Ti_3C_2T_x$ at 30 days, and

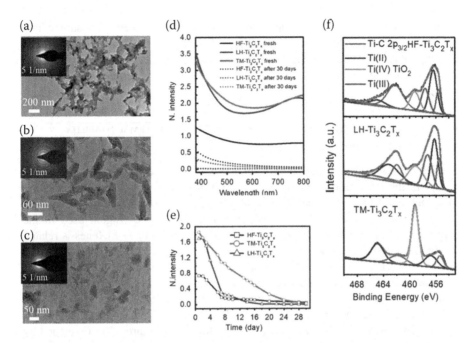

FIGURE 4.5 The TEM images of (a) HF-Ti$_3$C$_2$T$_x$, (b) LH-Ti$_3$C$_2$T$_x$, and (c) TM-Ti$_3$C$_2$T$_x$ suspensions, which are aged in the atmosphere at room temperature (RT) for 30 days. (d) The ultraviolet-visible spectrum of Ti$_3$C$_2$T$_x$ suspensions before and after 30 days. (e) The change in intensity of UV absorbance bands of Ti$_3$C$_2$T$_x$ suspensions during 30 days. (f) The XPS Ti 2p spectra of Ti$_3$C$_2$T$_x$ films after 30 days (Yang et al. 2020).

TM-Ti$_3$C$_2$T$_x$ suspension at 30 days. Flake and particle decomposition of Ti$_3$C$_2$T$_x$ was observed after aging. Among them, the particles in fresh HF-Ti$_3$C$_2$T$_x$ were degraded from a few microns to about 100 nm, and the slices in fresh LH-Ti$_3$C$_2$T$_x$ with a size of 200–300 nm or in fresh TM-Ti$_3$C$_2$T$_x$ with a size of several microns were degraded to tens of nanometers. The change of chemical composition in the 30-day sample was characterized by selective regional electron diffraction (SAED). The SAED diagram of the 30 day-HF-Ti$_3$C$_2$T$_x$ is identical to that of the 30 day-LH-Ti$_3$C$_2$T$_x$, corresponding to quadrangular titanium dioxide, indicating an oxidation reaction occurred to the sample. In contrast, the pattern of the 30 day-TM-Ti$_3$C$_2$T$_x$ corresponds to simple cubic aluminum hydroxide, which may be from the Al(OH)$_4^-$ surface terminal.

When stored in water, the –F group may be replaced by the –OH group and the -F terminating MXene is more unstable than the –O and/or –OH terminating groups. In the same work, it has also been reported that high temperature and/or metal absorption can cause the transformation from –OH group to –O group (Xie et al. 2014). Further studies showed that o-terminal MXenes could be decomposed into bare parts after contact with Mg, Ca, and Al. Naked MXene has been reported to be reactive in most cases. Exposed MXene is prone to oxidation in the presence of oxygen and water. This process leads to the formation of metal-oxide nanocrystals, which typically start at the edge of the sheet and eventually grow through the entire sheet. Exposure to light has also been reported to accelerate oxidation. Therefore, it is generally recommended that exposed MXene be stored in an oxygen-free, dry, and dark space (Mashtalir et al. 2014). Because of its instability, MXenes is functionalized by surface groups in most applications.

4.2.3 THE EFFECT OF FUNCTIONAL GROUPS ON THE THERMAL STABILITY OF MXENES

Studies have shown that the stability of MXenes is closely related to its surface group. In the preparation process of MXene in solution, surface groups including –F, –OH, and –O are attached to the surface. The type of surface group depends largely on the synthetic pathway and, in particular, on the reagent used in the etching. Experiments have shown that the oxidation stability of MXenes is related to etching conditions. For example, Feng et al. (Feng et al. 2017) successfully synthesized Ti_3C_2 by substituting NH_4HF_2 for HF, and clarified the influence of time, temperature, and Ti_3AlC_2 powder diameter on the etching process. The optimum etching time was 8 h, the temperature was 60 °C, and the average diameter of Ti_3AlC_2 was 325 mesh. Compared with HF etched Ti_3C_2, the thermal stability of Ti_3C_2 etched by NH_4HF_2 is better, and the multilayer structure of Ti_3C_2 etched by NH_4HF_2 can be retained even after 900 °C heat treatment. Thus, the thermal stability of MXenes can be improved by reducing defect density. The thermal stability of $Ti_3C_2T_x$ can be improved by using a low concentration of HF or by etching with mixed acids (such as sulfuric acid/HF or hydrochloric acid/HF) as compared to etching with a high concentration of HF.

For example, Seredych et al. (Seredych et al. 2019) used HF etching methods of different concentrations to explore their influence on the stability of MXenes (Figure 4.6). It can be seen that HF concentration increases from 5 wt% to 10 wt% (Figure 4.6b) or 30 wt% (Figure 4.6c), and substantial water loss can be observed (Figure 4.6d). Although each of the three samples had an H_2O peak at 320 °C, a new peak appeared at 200 °C with the increase of HF concentration, and the intensity of the bimodal peak increased with the increase of HF concentration. In addition, the initial temperature of released water decreased from 320 °C of 5 wt% HF sample to 200 °C of 10 wt% HF and 100 °C of 30 wt% HF sample. The shift to lower temperatures shows that the water content in $Ti_3C_2T_x$ increases with the increasing concentration of HF etchant. The MAX phase etched at a higher HF concentration leads to a more open structure in MXene than the MAX phase etched at a lower HF concentration. Compared with dense $Ti_3C_2T_x$ -5HF Mxene, $Ti_3C_2T_x$-30 HF Mxene has more macroscopic adsorption-layer water sites. As shown in Figure 4.6a–d, under the conditions of 5, 10, and 30 wt%, 0.35%, 1.9%, and 7.4% of total mass loss were directly related to outlet water at low temperature, respectively. As HF concentration increases, the relative amount of strongly bound water also increases. The increasing is due to an increase in surface imperfections, which causes water to bond with higher coordination bonds and thus release at higher temperatures. At temperatures above 400 °C, the weight loss of all three samples was similar.

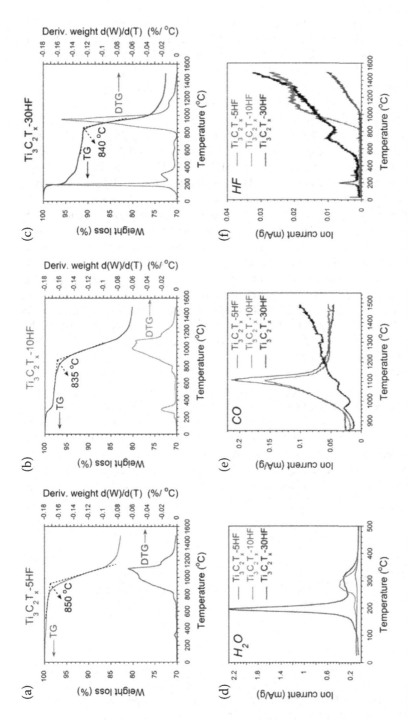

FIGURE 4.6 Thermal gravimetric (TG) curves with derivatives of the weight losses (DTG) for Ti$_3$C$_2$T$_x$ MXene obtained by etching Ti$_3$AlC$_2$ using three different HF concentrations of 5, 10, and 30 wt%: (a) Ti$_3$C$_2$T$_x$-5HF (5 wt% HF for etching), (b) Ti$_3$C$_2$T$_x$-10HF (10 wt% HF for etching), and (c) Ti$_3$C$_2$T$_x$-30HF (30 wt% HF for etching). Mass-spectrometry analysis for the atomic mass unit (amu) of (d) 18/H$_2$O, (e) 28/CO, and (f) 20/HF (Seredych et al. 2019).

$Ti_3C_2T_x$ remained stable at about 800 °C regardless of HF concentration, but the samples were degraded above this point (Figure 4.6). In all three samples (Figure 4.6e), CO was released at > 830 °C. Compared with 10 and 30 wt% HF, $Ti_3C_2T_x$-5HF showed the maximum CO release. On the contrary (Figure 4.6f), the weight loss caused by HF increases with the increase of HF etching concentration. These trends are due to the relative number of oxygen-group and fluorine-group terminals. Higher HF corrosion conditions increased F terminal. When the sample reaches 800 °C, it is gradually converted into cubic titanium carbide (TiC). At 1000 °C, a considerable amount of MXene remains, while at 1500 °C, the sample is completely converted into TiC. However, Raman spectroscopy showed that XRD amorphous TiO_2 and free carbon existed at 1000 °C. The former is reduced at 1500 °C to form cubic titanium carbide. The scanning electron microscope (SEM) of the initial ML-$Ti_3C_2T_x$ and the layered morphology of TiC tablets after heating to 1000 and 1500 °C under He.

Li et al. (Li et al. 2015) explained the oxidation reason. Ti_3C_2 2D thin sheet prepared in hydrogen fluoride solution has F and/or OH at its end. At the high temperature, terminated f and OH are lost or replaced by O. Therefore, the amount of terminated f and OH is the key factor in understanding TG and DSC curves. The EDS results show that the atomic ratio of Ti:C:F:O is 28:20:33:18. In previous studies, the ratio was 35:25:25:15, and the chemical formula was $Ti_{2.94}C_2F_2O_{0.55}(OH)_{0.65}$. According to EDS results and TG/DSC results (Figure 4.7), we propose a simple structure of MXene: $Ti_3C_2F_2OH$. As shown in Figure 4.7a, in the Ar atmosphere, the first-stage weightlessness (RT to 200 °C) is due to the loss of physically adsorbed water and HF on the surface of MXene. The second stage of weight loss (200–800 °C) is due to the loss of chemisorbed water, namely the hydroxyl group attached to the surface of Ti_3C_2. When the temperature exceeds 800 °C, the $Ti_3C_2F_2O_{0.5}$ obtained begins to lose F or O. Because of the bond energies, it should be easier to lose O than it is to lose F. Through analysis, it can be seen that the thermal process of MXene in Ar atmosphere is mainly a weightlessness process without obvious chemical reaction, which is confirmed by XRD and SEM. Weight loss requires energy to break the chemical bond between OH/O/F and the Ti_3C_2 crystal. So the corresponding DSC curve should be endothermic. However, as shown in Figure 4.7a, the true curve of

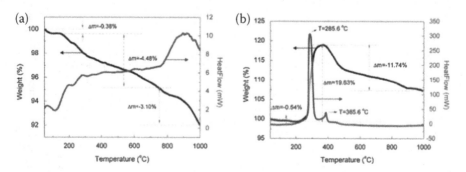

FIGURE 4.7 TG and DSC curves of MXene from room temperature to 1000 °C in (a) argon atmosphere and (b) oxygen atmosphere (Li et al. 2015).

DSC is endothermic only in the first stage, i.e., the loss of physically absorbed water. Subsequently, the DSC curve is completely transformed into exothermic heat, especially at the beginning of the second and third stages, related to the loss of chemisorbed water due to decreased surface energy. $Ti_3C_2F_2OH$ 2D thin plates have a large surface area and high surface energy.

They are thermodynamically metastable. When OH groups lose chemical bonds in the form of H_2O or O_2, some surfaces of MXene sheets change from Ti–OH to Ti–O or uncoordinated Ti. The newly formed surface is unstable and has high surface energy. In order to reduce the surface energy, the Ti–O on one wafer surface and the uncoordinated Ti on the other wafer surface attract each other so that the two slices attract each other and approach each other, forming a new metastable structure. Therefore, the two surfaces are transformed into a new interface. Some energy is released in the process, which explains the exothermic characteristics of the DSC curve in Figure 4.8a. However, compared with the peak in Figure 4.7b, the exothermic peak is very weak. The amount of heat released is very small. The new interface is very weak.

The high-temperature stability of MXenes is particularly important for its applications, especially in ceramics. At present, nearly 30 kinds of MXene have been reported through experiments, and dozens of them have been studied theoretically. Combining $4d$ or $5d$ transition metals and low atomic mass elements (i.e., B, N, and C) is of scientific importance in searching for high-temperature oxidation resistance, high hardness, and chemical resistance. Akgenc et al. (Akgenc 2019) determined that Ru_2C of Mo_2C, W_2C, Re_2C, Os_2C, and 2H phases with 1T and 2H phases might be synthesized in future experiments. AIMD calculation showed that 2H-W_2C, 1T-W_2C, 2H-MO_2C, and 2H-Re_2C MXene had good thermal stability and could keep their structure at 1500 K.

4.3 THE CHANGE OF PROPERTY OF MXENES ON THE PHOTOCATALYTIC PERFORMANCE OF MXENE-BASED PHOTOCATALYSTS

Theoretical calculation results show that the electronic structure of MXenes can be adjusted by adjusting the functional groups at the MXenes terminal (Tang, Zhou, and Shen 2012; Khazaei et al. 2015; Liu, Xiao, and Goddard 2016; Tahini, Tan, and Smith 2017). For example, the energy-band structure of the single primitive Ti_3C_2 layer is similar to that of a typical semimetal with finite-state density at the Fermi level. A clear separation between the conduction and valence bands is observed when its surface terminates at –F. Thus the band structure has semiconductor characteristics (Naguib et al. 2011). This happens because the band structure of MXenes varies significantly with the change of surface-functional groups, which will affect the role of MXene in receiving semiconductor charge carriers.

For example, ternary $Cu/TiO_2@Ti_3C_2T_x$ composite material was prepared by partial hydrothermal oxidation of $Ti_3C_2T_x$ and then light deposition (Peng et al. 2018). Scanning electron microscopy (STEM) and element mapping results in Figure 4.8d demonstrate the distribution of Cu species. The layered substrate at the lower-right corner can be attributed to $Ti_3C_2T_x$, and the nano-flakes are anatase TiO_2. The signal of the Cu element is highly overlapping with Ti and O, indicating that Cu is mostly

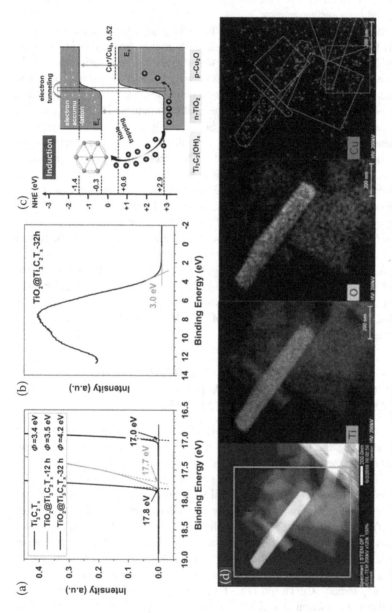

FIGURE 4.8 (a) UPS spectra of $Ti_3C_2T_x$, $TiO_2@Ti_3C_2T_x$ -12 h and $TiO_2@Ti_3C_2T_x$ -32 h and (b) valence band spectrum of $TiO_2@Ti_3C_2T_x$ -32 h, (c) shows the charge transfer at the interface of $Cu/TiO_2@Ti_3C_2(OH)_x$ during the induction period of water splitting reaction, (d) STEM image and EDS elemental mapping of $Cu/TiO_2@Ti_3C_2T_x$ -12 h (Peng et al. 2018).

deposited on the TiO_2 nanosheet but less on $Ti_3C_2T_x$. Consider that Cu(II) ions are reduced by the photoinduced electrons during the photo-deposition process. Therefore, the results show that the electrons generated by light are mainly left on TiO_2 for photoreduction Cu(II) rather than transferred to $Ti_3C_2T_x$. In other words, the $Ti_3C_2T_x$ sheet does not act as an electronic medium in this system.

In order to further confirm the charge transfer between $Ti_3C_2T_x$ and TiO_2, their electronic properties were studied. The result of ultraviolet photoelectron spectroscopy (UPS) is shown in Figure 4.8a, and the secondary electron cutoff of $Ti_3C_2T_x$ is 17.8 eV. On this basis, we calculated that the work function of $Ti_3C_2T_x$ is 3.4 eV (Peng et al. 2018). Under the hydrothermal reaction time of 32 h ($TiO_2@Ti_3C_2T_x$-32 h), the work function of $TiO_2@Ti_3C_2T_x$ is 4.2 eV, and the valence band is 3.0 eV (Figure 4.8b), both of which are quite close to the reported values of anatase TiO_2 (Xiong et al. 2007). The results indicate that $TiO_2@Ti_3C_2T_x$-32 h is mainly composed of TiO_2, and $Ti_3C_2T_x$ is almost exhausted. Therefore, in the prepared $TiO_2@Ti_3C_2T_x$ composite, the work function (3.4 eV) of $Ti_3C_2T_x$ is lower than that of TiO_2 (4.2 eV), which enables holes to flow from TiO_2 (high work-function material) to $Ti_3C_2T_x$ (low work-function material) through the interface, as shown in Figure 4.8c.

In addition, the work function of the hybrid $TiO_2@Ti_3C_2T_x$-12 h is 3.5 eV, which is between TiO_2 and $Ti_3C_2T_x$, further proving that $Ti_3C_2T_x$ MXene has the potential as a hole trap with low work function and can change the electronic performance of TiO_2 photocatalyst. Compared with $Cu/TiO_2@Ti_3C_2T_x$-32h, the enhanced photo-activity of $Cu/TiO_2@Ti_3C_2T_x$-12 h is beneficial to hydrogen production under simulated sunlight (Peng et al. 2018). The function of MXenes as hole acceptors has also been reported in other MXenes-TiO_2 composites. As mentioned above, MXenes have surface metal sites and functional termination functional groups (e.g., –OH, –O, and –F). These surface materials have been shown to act as active centers for electrocatalytic applications, such as hydrogen evolution reactions (Wang et al. 2018; Li and Wu 2019). In recent years, researchers have begun to exploit MXenes with specific surface properties as active sites for photocatalytic reactions (Sun et al. 2018; Gao, O'Mullane, and Du 2017; Ye et al. 2018).

In recent years, researchers have moved to exploit MXenes to have specific surface properties as active sites for photocatalytic reactions. Generally, the overall H_2 precipitation reaction (HER) approach can be summarized as the formation of Initial State $H^+ + e^-$, intermediate adsorption H^*, and final product $1/2H_2$ (Jiao et al. 2015). The intermediate state of Gibbs free energy $|\Delta G_{H^*}|$ is regarded as important an index of various kinds of ehrs activity of catalysts. ΔG_{H^*} of the ideal value should be zero (Jiao et al. 2015). The potential of $Ti_3C_2T_x$ as a catalyst for hydrogen evolution was discussed by DFT calculation (Ran et al. 2017). The Gibbs free energy of pure Ti_3C_2, O terminal Ti_3C_2, and F terminal Ti_3C_2 surface atom adsorption H, $|\Delta G_{H^*}|$ is calculated as the main index of hydrogen production activity of catalytic reaction. The results (Figure 4.9a) show that pure Ti_3C_2 presents a negative $|\Delta G_{H^*}|$ at the optimal H^* coverage range (= 1/8), with a value of −1.021 eV, indicating that H^* has strong chemical adsorption on its surface. For Ti_3C_2 at the F end, positive $|\Delta G_{H^*}|$ can be observed to be 1.995 eV (Figure 4.9b), which indicates that the adsorption of H^* is very weak, enabling an easy desorption of products. Pure Ti_3C_2 and F-terminal Ti_3C_2 are unfavorable to hydrogen-evolution reaction

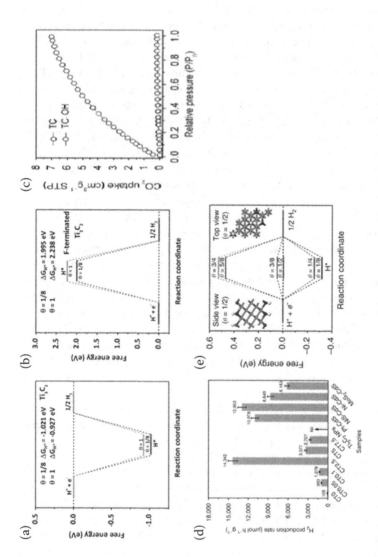

FIGURE 4.9 (a) and (b) show the calculated Gibbs free energies for H adsorption on the surface of a $2 \times 2 \times 1$ Ti_3C_2 supercell with different H*coverages, respectively (Ran et al. 2017), (c) CO_2 adsorption behaviors for TC and TC-OH, (d) A comparison of the photocatalytic H_2-production activities of CT0, CT0.05, CT0.1, CT2.5, CT5, CT7.5, Ti_3C_2 NPs, Pt-CdS, NiS-CdS, Ni-CdS and MoS_2-CdS, (e) The calculated free-energy diagram of HER at the equilibrium potential ($U = 0$ V) on the surface of a $2 \times 2 \times 1$ O-terminated Ti_3C_2 supercell at different H* coverage (1/8, 1/4, 3/8, 1/2, 5/8, and 3/4) conditions (Ran et al. 2017).

(HER). It is worth noting that in the best H^* coverage ($\theta = 1/2$), the $|\Delta G_{H^*}|$ of O-terminated Ti_3C_2 is 0.00283 eV, which is close to zero (Ran et al. 2017), even lower than the value of noble metal Pt (0.09 eV) (Noerskov et al. 2005), and the sum of MoS_2 (0.08 eV) and WS_2 (0.08 eV) (Bonde et al. 2008)of highly active HER catalyst (Figure 4.9e). These results indicate that O-terminated Ti_3C_2 is beneficial as a co-catalyst for hydrogen production, also demonstrated by photoactivity studies. It can be seen from Figure 4.9d that the introduction of O-terminal Ti_3C_2 significantly improves the hydrogen-evolution activity of Ti_3C_2-CDS composite under visible light irradiation ($\geq$420 nm). The best performance was observed at 2.5 wt% Ti_3C_2-CDS, even better than usual co-catalysts (such as Pt, NiS, Ni, and MoS_2) (Figure 4.9d). These results suggest that O-terminated Ti_3C_2 may be a promising alternative to noble metal Pt, thus providing an active site for hydrogen evolution.

In addition, MXene with specific surface-functional groups has also been used as an active site for photocatalytic reduction of CO_2 (Zhang, Zhang, et al. 2017; Ye et al. 2018). The hydroxyl groups of MXenes can provide the adsorption and activation of acidic CO_2. Ye et al. (Ye et al. 2018) used the KOH treatment method to regulate the functional groups on the surface of Ti_3C_2 by replacing –F with OH. Analysis of the adsorption of CO_2 by unmodified Ti_3C_2 (expressed as TC) and surface alkalization of Ti_3C_2 (expressed as TC-OH) shows that the absorption of CO_2 is TC-OH much higher than that of TC at room temperature (Figure 4.9c). Based on the calculation results of DFT, the adsorption energy (E_{ad}) of CO_2 on TC-OH is more negative than that of Ti_3C_2 terminated by F, indicating that CO_2 molecules are more easily adsorbed on the surface of TC-OH. For unused and used samples, fresh and used samples were prepared by mechanically mixing Ti_3C_2 material with commercially available P25, which was further confirmed by Fourier transform infrared spectroscopy (FTIR) results. The difference in TC/P25 absorption bands before and after the reaction was negligible. In sharp contrast, strong absorption bands of various carbonate species, such as Vs (CO_3) ($\approx$1320 cm^{-1}) and Vas (CO_3) ($\approx$1580 cm^{-1}), appeared on TC-OH/P25 after irradiation, indicating that CO_2 adsorbed by TC-OH/P25 could be effectively converted into easily activated species (for example, CO_3^{2-}) (Ye et al. 2018). These results indicate that surface alkalinization makes $Ti_3C_2T_x$ having more surface hydroxyl groups that become active sites for the adsorption and activation of CO_2 molecules, thus contributing to the enhancement of photoactivity and the reduction of CO_2 into CO and CH_4 (Ye et al. 2018). In addition, the simulation results based on first-principles calculation show that Ti_2CO_2 of single-layer O terminal can be photocatalytic CO_2 reduction, in which oxygen vacancy acts as the active site for CO_2 adsorption and activation (Zhang, Zhang, et al. 2017). These reports demonstrate the great potential and rich tunability of MXene to provide active sites for various photocatalytic reactions.

ACKNOWLEDGMENTS

This work was supported by the National Natural Science Foundation of China (22078074, 21938001), Guangxi Natural Science Foundation (2019GXNSFAA245006, 2020GXNSFDA297007, 2016GXNSFFA380015), Special funding for 'Guangxi Bagui

Scholars', and Scientific Research Foundation for High-level Personnel from Guangxi University.

REFERENCES

Akgenc, B. 2019. New Predicted Two-Dimensional MXenes and Their Structural, Electronic and Lattice Dynamical Properties. *Solid State Communications* 303:7.

Anasori, B., Y. Xie, M. Beidaghi, et al. 2015. Two-Dimensional, Ordered, Double Transition Metals Carbides (MXenes). *ACS Nano* 9 (10):9507–9516.

Anasori, Babak, Maria R. Luhatskaya, and Yury Gogotsi. 2017. 2D Metal Carbides and Nitrides (MXenes) for Energy Storage. *Nature Reviews Materials* 2:16098.

Anasori, Babak, Chenyang Shi, Eun Ju Moon, et al. 2016. Control of Electronic Properties of 2D Carbides (MXenes) by Manipulating Their Transition Metal Layers. *Nanoscale Horizons* 1 (3):227–234.

Bonde, J., P. G. Moses, T. F. Jaramillo, J. K. Norskov, and I. Chorkendorff. 2008. Hydrogen Evolution on Nano-Particulate Transition Metal Sulfides. *Faraday Discussions* 140:219–231.

Caffrey, Nuala M. 2018. Effect of Mixed Surface Terminations on the Structural and Electrochemical Properties of Two-Dimensional $Ti_3C_2T_x$ and V_2CT_2 MXenes Multilayers. *Nanoscale* 10 (28):13520–13530.

Feng, Aihu, Yun Yu, Feng Jiang, et al. 2017. Fabrication and Thermal Stability of NH_4HF_2-Etched Ti_3C_2 MXene. *Ceramics International* 43 (8):6322–6328.

Gandi, Appala Naidu, Husam N. Alshareef, and Udo Schwingenschloegl. 2016. Thermoelectric Performance of the MXenes M_2CO_2 (M = Ti, Zr, or Hf). *Chemistry of Materials* 28 (6):1647–1652.

Gao, Guoping, Anthony P. O'Mullane, and Aijun Du. 2017. 2D MXenes: A New Family of Promising Catalysts for the Hydrogen Evolution Reaction. *ACS Catalysis* 7 (1):494–500.

Hope, Michael A., Alexander C. Forse, Kent J. Griffith, et al. 2016. NMR Reveals the Surface Functionalisation of Ti_3C_2 MXene. *Physical Chemistry Chemical Physics* 18 (7):5099–5102.

Hu, M., Z. Li, T. Hu, S. Zhu, C. Zhang, and X. Wang. 2016. High-Capacitance Mechanism for $Ti_3C_2T_x$ MXene by in Situ Electrochemical Raman Spectroscopy Investigation. *ACS Nano* 10 (12):11344–11350.

Hu, Tao, Jiemin Wang, Hui Zhang, Zhaojin Li, Minmin Hu, and Xiaohui Wang. 2015. Vibrational Properties of Ti_3C_2 and $Ti_3C_2T_2$ (T = O, F, OH) Monosheets by First-Principles Calculations: A Comparative Study. *Physical Chemistry Chemical Physics* 17 (15):9997–10003.

Jiao, Yan, Yao Zheng, Mietek Jaroniec, and Shi Zhang Qiao. 2015. Design of Electrocatalysts for Oxygen- and Hydrogen-Involving Energy Conversion Reactions. *Chemical Society Reviews* 44 (8):2060–2086.

Kajiyama, Satoshi, Lucie Szabova, Keitaro Sodeyama, et al. 2016. Sodium-Ion Intercalation Mechanism in MXene Nanosheets. *Acs Nano* 10 (3):3334–3341.

Khazaei, M., M. Arai, T. Sasaki, et al. 2013. Novel Electronic and Magnetic Properties of Two-Dimensional Transition Metal Carbides and Nitrides. *Advanced Functional Materials* 23 (17):2185–2192.

Khazaei, Mohammad, Masao Arai, Taizo Sasaki, Ahmad Ranjbar, Yunye Liang, and Seiji Yunoki. 2015. OH-Terminated Two-Dimensional Transition Metal Carbides and Nitrides as Ultralow Work Function Materials. *Physical Review B* 92 (7):075411.

Khazaei, Mohammad, Ahmad Ranjbar, Masao Arai, Taizo Sasaki, and Seiji Yunoki. 2017. Electronic Properties and Applications of MXenes: A Theoretical Review. *Journal of Materials Chemistry C* 5 (10):2488–2503.

Kim, Hyunho, Zhenwei Wang, and Husam N. Alshareef. 2019. MXetronics: Electronic and Photonic Applications of MXenes. *Nano Energy* 60:179–197.

Lee, Youngbin, Sung Beom Cho, and Yong-Chae Chung. 2014. Tunable Indirect to Direct Band Gap Transition of Monolayer Sc_2CO_2 by the Strain Effect. *Acs Applied Materials & Interfaces* 6 (16):14724–14728.

Lee, Youngbin, Yubin Hwang, Sung Beom Cho, and Yong-Chae Chung. 2014. Achieving a Direct Band Gap in Oxygen Functionalized-Monolayer Scandium Carbide by Applying an Electric Field. *Physical Chemistry Chemical Physics* 16 (47):26273–26278.

Li, Longhua. 2016. Effects of the Interlayer Interaction and Electric Field on the Band Gap of Polar Bilayers: A Case Study of Sc_2CO_2. *Journal of Physical Chemistry C* 120 (43):24857–24865.

Li, Xiuqin, Chengyin Wang, Yu Cao, and Guoxiu Wang. 2018. Functional MXene Materials: Progress of Their Applications. *Chemistry-An Asian Journal* 13 (19):2742–2757.

Li, Z. and Y. Wu. 2019. 2D Early Transition Metal Carbides (MXenes) for Catalysis. *Small* 15 (29):1804736.

Li, Zhengyang, Libo Wang, Dandan Sun, et al. 2015. Synthesis and Thermal Stability of Two-Dimensional Carbide MXene Ti_3C_2. *Materials Science and Engineering B-Advanced Functional Solid-State Materials* 191:33–40.

Lipatov, Alexey, Mohamed Alhabeb, Maria R. Lukatskaya, Alex Boson, Yury Gogotsi, and Alexander Sinitskii. 2016. Effect of Synthesis on Quality, Electronic Properties and Environmental Stability of Individual Monolayer Ti_3C_2 MXene Flakes. *Advanced Electronic Materials* 2 (12):1600255.

Liu, Yuan, Jian Guo, Enbo Zhu, et al. 2018. Approaching the Schottky-Mott Limit in Van der Waals Metal-Semiconductor Junctions. *Nature* 557 (7707):696–700.

Liu, Yuanyue, Paul Stradins, and Su-Huai Wei. 2016. Van der Waals Metal-Semiconductor Junction: Weak Fermi Level Pinning Enables Effective Tuning of Schottky Barrier. *Science Advances* 2 (4): e1600069.

Liu, Yuanyue, Hai Xiao, and William A. Goddard, III. 2016. Schottky-Barrier-Free Contacts with Two-Dimensional Semiconductors by Surface-Engineered MXenes. *Journal of the American Chemical Society* 138 (49):15853–15856.

Lukatskaya, M. R., O. Mashtalir, C. E. Ren, et al. 2013. Cation Intercalation and High Volumetric Capacitance of Two-Dimensional Titanium Carbide. *Science* 341 (6153):1502–1505.

Luo, J., X. Tao, J. Zhang, et al. 2016. Sn^{4+} Ion Decorated Highly Conductive Ti_3C_2 MXene: Promising Lithium-Ion Anodes with Enhanced Volumetric Capacity and Cyclic Performance. *ACS Nano* 10 (2):2491–2499.

Magne, Damien, Vincent Mauchamp, Stéphane Célérier, Patrick Chartier, and Thierry Cabioc'H. 2015. Spectroscopic Evidence in the Visible-Ultraviolet Energy Range of Surface Functionalization Sites in the Multilayer Ti_3C_2 MXene. *Physical Review B* 91 (20):201409–201414.

Mashtalir, O., K. M. Cook, V. N. Mochalin, M. Crowe, M. W. Barsoum, and Y. Gogotsi. 2014. Dye Adsorption and Decomposition on Two-Dimensional Titanium Carbide in Aqueous Media. *Journal of Materials Chemistry A* 2 (35):14334–14338.

Mashtalir, O., M. R. Lukatskaya, A. I. Kolesnikov, et al. 2016. The Effect of Hydrazine Intercalation on the Structure and Capacitance of 2D Titanium Carbide (MXene). *Nanoscale* 8 (17):9128–9133.

Mohammad Khazaei, Ahmad Ranjbar, Masao Arai, and Seiji Yunoki. 2016. Topological Insulators in the Ordered Double Transition Metals $M'_2M''C_2$ MXenes (M' = Mo, W; M'' = Ti, Zr, Hf). *Physical Review B* 94:125152.

Naguib, M., M. Kurtoglu, V. Presser, et al. 2011. Two-Dimensional Nanocrystals Produced by Exfoliation of Ti_3AlC_2. *Advanced Materials* 23 (37):4248–4253.

Noerskov, J. K., T. Bligaard, A. Logadottir, et al. 2005. Trends in the Exchange Current for Hydrogen Evolution. *Cheminform* 36 (24):e12154–e12154.

Peng, Chao, Ping Wei, Xiaoyao Li, Yunpeng Liu, and Kangle Lv. 2018. High Efficiency Photocatalytic Hydrogen Production over Ternary $Cu/TiO_2@Ti_3C_2T_x$ Enabled by Low-Work-Function 2D Titanium Carbide. *Nano Energy* 53:97–107.

Ran, Jingrun, Guoping Gao, Fa-Tang Li, Tian-Yi Ma, Aijun Du, and Shi-Zhang Qiao. 2017. Ti_3C_2 MXene Co-Catalyst on Metal Sulfide Photo-Absorbers for Enhanced Visible-Light Photocatalytic Hydrogen Production. *Nature Communications* 8:13907.

Seh, Z. W., K. D. Fredrickson, B. Anasori, et al. 2016. Two-Dimensional Molybdenum Carbide (MXene) as an Efficient Electrocatalyst for Hydrogen Evolution. *ACS Energy Letters* 1 (3):589–594.

Seredych, Mykola, Christopher Eugene Shuck, David Pinto, et al. 2019. High-Temperature Behavior and Surface Chemistry of Carbide MXenes Studied by Thermal Analysis. *Chemistry of Materials* 31 (9):3324–3332.

Si, Chen, Kyung-Hwan Jin, Jian Zhou, Zhimei Sun, and Feng Liu. 2016. Large-Gap Quantum Spin Hall State in MXenes: D-Band Topological Order in a Triangular Lattice. *Nano Letters* 16 (10):6584–6591.

Sun, Yuliang, Di Jin, Yuan Sun, et al. 2018. g-$C_3N_4/Ti_3C_2T_x$ (MXenes) Composite with Oxidized Surface Groups for Efficient Photocatalytic Hydrogen Evolution. *Journal of Materials Chemistry A* 6 (19):9124–9131.

Tahini, Hassan A., Xin Tan, and Sean C. Smith. 2017. The Origin of Low Workfunctions in OH Terminated MXenes. *Nanoscale* 9 (21):7016–7020.

Tang, Qing, Zhen Zhou, and Panwen Shen. 2012. Are MXenes Promising Anode Materials for Li Ion Batteries? Computational Studies on Electronic Properties and Li Storage Capability of Ti_3C_2 and $Ti_3C_2X_2$ (X = F, OH) Monolayer. *Journal of the American Chemical Society* 134 (40):16909–16916.

Wang, Hou, Yan Wu, Xingzhong Yuan, et al. 2018. Clay-Inspired MXene-Based Electrochemical Devices and Photo-Electrocatalyst: State-of-the-Art Progresses and Challenges. *Advanced Materials* 30 (12):1704561.

Weng, Hongming, Ahmad Ranjbar, Yunye Liang, et al. 2015. Large-Gap Two-Dimensional Topological Insulator in Oxygen Functionalized MXene. *Physical Review B* 92 (7):075436.

Xie, Yu, and P. R. C. Kent. 2013. Hybrid Density Functional Study of Structural and Electronic Properties of Functionalized $Ti_{n+1}X_n$(X = C, N) Monolayers. *Physical Review B* 87:235441.

Xie, Yu, Michael Naguib, Vadym N. Mochalin, et al. 2014. Role of Surface Structure on Li-Ion Energy Storage Capacity of Two-Dimensional Transition-Metal Carbides. *Journal of the American Chemical Society* 136 (17):6385–6394.

Xiong, Gang, Rui Shao, Timothy C. Droubay, et al. 2007. Photoemission Electron Microscopy of TiO_2 Anatase Films Embedded with Rutile Nanocrystals. *Advanced Functional Materials* 17 (13):2133–2138.

Xu, Jiao, Jaewoo Shim, Jin-Hong Park, and Sungjoo Lee. 2016. MXene Electrode for the Integration of WSe_2 and MoS_2 Field Effect Transistors. *Advanced Functional Materials* 26 (29):5328–5334.

Yang, Yina, Zherui Cao, Liangjing Shi, Ranran Wang, and Jing Sun. 2020. Enhancing the Conductivity, Stability and Flexibility of $Ti_3C_2T_x$ MXenes by Regulating Etching Conditions. *Applied Surface Science* 533:147475.

Ye, Minheng, Xin Wang, Enzuo Liu, Jinhua Ye, and Defa Wang. 2018. Boosting the Photocatalytic Activity of P25 for Carbon Dioxide Reduction by using a Surface-Alkalinized Titanium Carbide MXene as Cocatalyst. *Chemsuschem* 11 (10):1606–1611.

Yu, Xue Fang, Jian Bo Cheng, Zhen Bo Liu, et al. 2015. The Band Gap Modulation of Monolayer Ti_2CO_2 by Strain. *RSC Advances* 5 (39):30438–30444.

Zha, Xian-Hu, Qing Huang, Jian He, et al. 2016. The Thermal and Electrical Properties of the Promising Semiconductor MXene Hf_2CO_2. *Scientific Reports* 6:27971.

Zhang, Chuanfang John, Sergio Pinilla, Niall McEyoy, et al. 2017. Oxidation Stability of Colloidal Two-Dimensional Titanium Carbides (MXenes). *Chemistry of Materials* 29 (11):4848–4856.

Zhang, Xu, Zihe Zhang, Jielan Li, Xudong Zhao, Dihua Wu, and Zhen Zhou. 2017. Ti_2CO_2 MXene: A Highly Active and Selective Photocatalyst for CO_2 Reduction. *Journal of Materials Chemistry A* 5:12899–12903.

5 Construction of MXene-Based Photocatalysts

Sze-Mun Lam, Zi-Jun Yong, Man-Kit Choong,
Jin-Chung Sin, Ying-Hui Chin, and Jin-Han Tan
Department of Environmental Engineering, Faculty of
Engineering and Green Technology, Universiti Tunku
Abdul Rahman

CONTENTS

5.1 INTRODUCTION

Heterogeneous photocatalysis is an innovative catalysis technology for transforming solar energy into chemical energy, which has demonstrated huge potential in alleviating the thorny problems of the energy crisis and environmental pollution. In the photocatalysis mechanism, it commonly involved: (1) absorption of adequate light energy by the photocatalysts to generate electron-hole pairs, (2) separation and migration of charge carriers to the surface of photocatalyst, and (3) surface-redox reactions by consuming the separated charge carriers (Baral, Reddy, and Parida 2019; Chen et al. 2019; Gopinath et al. 2020). Among them, the charge transfer and separation considered the rate-determining step in the photoreaction because the slow migration and fast recombination of charge carriers seriously inhibited the photoactivity and solar conversion efficacy. Therefore, a lot of approaches have been taken with the eye to accelerate the charge separation efficiency for obtaining effective photocatalytic applications.

DOI: 10.1201/9781003156963-5

The benefits originated from MXene as photocatalysts have triggered much research in this field (Guo et al. 2016; Zhao, Nivetha, et al. 2020; Feng et al. 2021). MXene has been a novel guaranteeing series of 2D catalyst with incredible features such as high surface area, superb hydrophilic surface, great metallic and electrical conductivity, and excellent oxidation resistance. Since the revolutionary work by researchers at Drexel University on the enhancement of Ti_3C_2 MXene through wet-chemical etching in HF acid and exfoliation of Ti_3AlC_2, various types of MXenes have been developed via several methods for varied photocatalytic usages (Low et al. 2018; Cheng et al. 2019; Kuang et al. 2020; Liu and Chen 2020; Zhao, Nivetha, et al. 2020). MXenes itself is a versatile material found applications in organic pollutants photodegradation, CO_2 photoconversion, heavy-metal photoreduction, H_2 photo evolution, nitrogen fixation, and even in bacteria photoinactivation (Low et al. 2018; Hao et al. 2020; Huang, Li, and Meng 2020; Liu and Chen 2020; Huang et al. 2021; Lu et al. 2021).

Heterostructure photocatalysts containing MXenes were assembled extensively to promote the interfacial charge transfer and prevented the recombination of electron-hole pairs (Huang, Li, and Meng 2020; Liu and Chen 2020). In this chapter, the main focus is on the preparation of MXenes-based photocatalysts, with emphasis on the efficacy of the photocatalytic materials synthesized by different techniques in any considered photoactivity. This has been believed vital, and many current investigations in this area are underway. For the benefit of novice researchers, this chapter places formerly reported data available on the preparation of MXenes-based photocatalysts together with the merits and demerits so that novice researchers can save resources and time when they start their research studies.

5.2 PREPARATION OF MXENE-BASED PHOTOCATALYSTS AND EVALUATION OF THEIR PHOTOACTIVITY

Considering the huge increase in the utilization of MXene-based nanomaterials, numerous reviews on their preparation for different applications have emerged (Hong et al. 2020; Kuang et al. 2020; Im et al. 2021). Notably, MXene-based nanocomposite photocatalysts were fabricated by decorating with different semiconductors or support materials for better photoactivities (Alfahel et al. 2020; Du et al. 2020). The extensively utilized synthetic methods for preparing the MXene-based nanocomposites are mechanical mixing, ultrasonic self-assembly, in situ growth, and hydrothermal method. Mechanical mixing of different components in solution is the easiest method of fabricating the nanocomposite photocatalysts (Song and Zhang 2010). The ultrasonic self-assembly was carried out based on the electrostatic attraction between negatively charged MXene and positively charged photocatalysts (Zhang et al. 2020). Unlike both abovementioned methods, in situ growth can produce strong chemically bonded nanocomposites by directly growing different components on the surface of MXenes (Khan and Tahir 2021). In contrast, the hydrothermal technique can offer an affordable and convenient way of heat treatment that has frequently led to high yields of nanocomposite photocatalyst. The hydrothermal synthesis is also an eco-friendly soft-solution method since the process does not require the exploitation of harmful reactants and results in no emission of undesired pollutants (Liu, Yang, Jin, et al. 2020).

5.2.1 PREPARATION OF PRISTINE MXENE

Different synthetic approaches have been used to improve the photoactivity of pristine MXene, including morphology thickness control and alteration of catalyst colloid concentration. Zhao et al. (Zhao, Que, et al. 2020) reported that MXene was synthesized by selecting an appropriate etchant to break down the weak binding force between A sheet and MX sheet in the MAX phase, thus producing layered materials with alternating M and X phases. In addition, the MXene nanosheets with few or single-layers were obtained by intercalation and ultrasonic removal. Meanwhile, the functional groups of MXene, such as -OH, -F, -O, etc., were generated by applying an appropriate etchant (eg. NaOH, LiF+HCl, HF). In general, there are three main etching approaches: fluorinated etching, fluorine-free etching, and electrochemical etching.

Using the fluorinated-etchant method, Liu et al. (Liu, Yang, Ren, et al. 2020) obtained the 2D Ti_3C_2 by etching the Ti_3AlC_2 powder with HF acid. The Ti_3AlC_2 was synthesized by mixing the Ti_2AlC and TiC in a ratio of 1:1 and ball milled for 24 h. The corundum boat coated with 5 g NaCl was then prepared, and the well-mixed mixture was loaded on and heated to 1150°C at a heating rate of 5°C min^{-1} and held for 2 h under an argon atmosphere in a tube furnace. The resulting products were washed and stirred in 30 mL of 9 mol L^{-1} HCl solution for 12 h at 85°C. The Ti_3AlC_2 obtained through filtration and vacuum dried at 60°C. To fabricate 2D Ti_3C_2, 1 g of synthesized Ti_3AlC_2 was added slowly and carefully into 30 mL of concentrated HF (40%), thus to avoid a violent reaction. The suspension was stirred for 24 h at room temperature, washed with deionized water, and centrifuged at 3500 rpm several times until the pH of the supernatant adjusted back to pH 7.0. The obtained 2D Ti_3C_2 powder was finally dried under vacuum conditions at 60°C for 12 h inside a vacuum oven. The whole synthetic procedure for preparing the Ti_3C_2 is schematically shown in Figure 5.1.

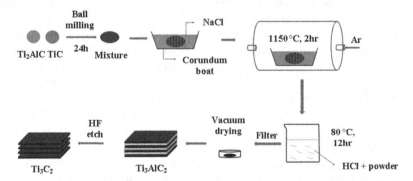

FIGURE 5.1 Schematic diagram of the NaCl-aided preparation of Ti_3AlC_2 from Ti_2AlC-TiC mixture and the subsequent production of Ti_3C_2 (Liu, Yang, Ren, et al. 2020).

Li et al. (Li, Han, et al. 2021) reported that Ti2C MXene was fabricated through an etching-exfoliating method in a solution of LiF and HCl to remove Al atom from Ti_2AlC. Researchers prepared 6 M of HCl solution by mixing 15 mL of 12 M HCl solution with 15 mL of distilled water in a polytetrafluoroethylene conical bottle. After being stirred and heated for 10 min, 1.98 g LiF powder was added into a diluted HCl solution to prepare LiF/HCl solution. The LiF/HCl solution was heated to 80 °C, and then 3 g Ti_2AlC was slowly and carefully added into the polytetrafluoroethylene bottle. The mixture was stirred, centrifuged, and washed thoroughly with ethanol, as well as water several times, to adjust the pH closer to neutrality. The obtained precursor was then dried, dispersed into 100 mL of dimethyl sulphoxide (DMSO), and stirred for 24 h at room temperature. Subsequently, the solid was obtained by centrifuging and washing with distilled water. The solid was then dispersed into distilled water and sonicated for 1 h. The Ti_2C MXene was collected by centrifuging at 3500 rpm for 1 h and finally vacuum freeze-drying at 350 °C for 8 h.

In 2020, Natu et al. (Natu et al. 2020) successfully synthesized the 2D Ti_3C_2 MXene by etching the MAX phase in polar organic solvents, as displayed in Figure 5.2 At first, Ti_3AlC_2 powders were made by mixing titanium carbide, Al, and Ti powders with a molar ratio of 2:1.05:1, respectively. The powders were then ball milled at 70 rpm for 24 h. Ball-milled powders transferred to an alumina boat, which was placed inside an alumina tube furnace and heated with flowing of Ar (15 SSCM) at 1450 °C for 2 h. The heating and cooling rates were set at 5 °C min^{-1}. After sieving with a 400 μm mesh, the Ti_3AlC_2 powders were added to 10 mL of propylene carbonate that premixed with 1 g dehydrated NH_4HF_2 and stirred at 500 rpm inside an Ar-filled glove box for 196 h at 35°C. The resulted slurry was washed with acidic propanol solution and propylene carbonate for several times. For delamination, 30 mL of propylene carbonate was added to $Ti_3C_2T_z$ sediments and was sonicated under flowing Ar for 1 h in an ice bath. The suspension was then centrifuged to collect the delaminated $Ti_3C_2T_z$ from the insoluble salts produced during etching and multilayers that were not delaminated. The supernatant $Ti_3C_2T_z$ colloid was filtered via a membrane with the assistance of a vacuum pump. The filtered film was vacuum dried and ground with mortar and pestle.

MXene synthesized by fluorine-free route was reported by Mei et al. (Mei et al. 2020). The synthesis route is illustrated in Figure 5.3a. High purity of Mo_2GaC was needed to synthesize the 2D mesoporous Mo_2C. Approximately 700 mg of Mo_2GaC was mixed with 50 mL of H_3PO_4 acid. The mixture was then magnetically stirred at 1000 rpm for 4 h under UV irradiation. The resulting mixture was centrifuged and washed with deionized water several times to adjust the pH of the supernatant close to 7.0. The obtained powder was also re-dispersed into oxygen-free deionized water and ultrasonically treated in an ice water bath. Lastly, the mesoporous Mo_2C MXene powder was fabricated after a series of centrifuging and vacuum drying at 60°C. Their research demonstrated that the fabrication of fluorine-free Mo_2C was obtained through a UV-induced selective-etching method with a Mo_2GaC bulk precursor. They went further to explain that this synthetic method was an eco-friendly route since no hazardous and highly corrosive acids were used. As shown in Figure 5.3b, the Mo_2Ga_2C powder was a typical platelet-like morphology with lateral size of ~5 μm and thickness of ~1 μm. An obvious layered structure was observed after selective

(a)

NH₄HF₂ Propylene
Carbonate 35 °C, 168 h   Etching out Al which
(Ar)   forms AIF₃ + (NH₄)₃AIF₆

(b)

HCl / PrOH Washing

(c)

Vacuum filtering after dispersing
MXene in Propylene Carbonate

(d)

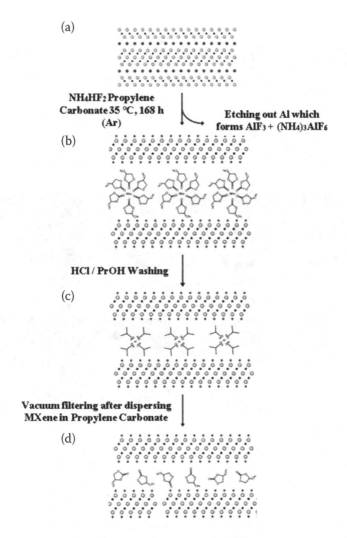

FIGURE 5.2 Schematic illustration of etching and washing process in fabricating the Ti₃C₂Tz (Natu et al. 2020).

etching with UV irradiation in H₃PO₄ acid from Figure 5.3c, d. The etched sample was further revealed in Figure 5.3e by field-emission scanning electron microscope and showed a planer size of several mm with thickness of few hundred nm. Energy dispersive X-ray (EDX) mapping in Figure 5.3f–h verified the presence of Mo and C elements, while O was detected due to the partial oxidation of surface-active edges or the residual oxygen-containing group.

Li et al. (Li et al. 2019) described a two-step alkali route to synthesize Ti₃C₂Tx nanosheets, as shown in Figure 5.4. At the beginning of the process, the Ti₃AlC₂ powder was treated with 27.5 M NaOH at 270 °C under an argon gas atmosphere. The Ti₃C₂Tx flakes were then generated by tetramethylammonium hydroxide (TMAOH) intercalation for 24 h. The mixture was sonicated for 6 h under an Ar

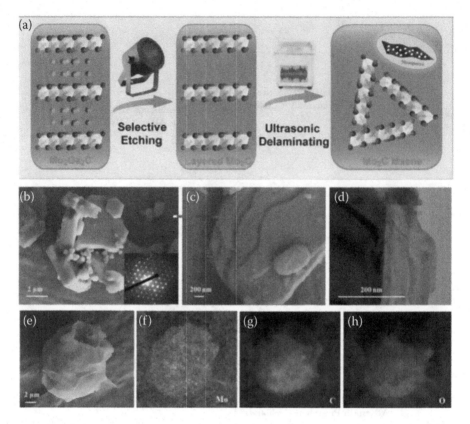

FIGURE 5.3 (a) Synthetic route of UV-induced selective etching of Mo_2C. FESEM images of (b) Mo_2GaC powder and (c) Mo_2Ga_2C after UV-induced selective etching. (d) TEM image of Mo_2Ga_2C after UV-induced selective etching. (e) FESEM image of layered Mo_2C and the EDX mapping of (f) Mo, (g) C, and (h) O elements (Mei et al. 2020).

atmosphere. After that, the solution was centrifuged at 10000 rpm for 30 min, and the supernatant was transferred and preserved for further experiments. With their proposed method, a smaller Ti_3C_2 was produced, and the performance was about twice as effective as compared to the samples obtained via fluorine-based treatment.

5.2.2 Preparation of Binary MXene-Based Nanocomposites

The unique 2D sheet morphology of MXene can serve as an excellent platform for the heterojunction construction with other semiconductors to increase its photoactivity. For example, in the presence of light irradiation, the binary MXene-based nanocomposites can accelerate the electron-hole pair separation and increase the light-absorption ability. Hence, the binary MXene-based nanocomposites were preferred over the pristine MXene in various photocatalytic applications. Table 5.1 summarizes the recent development of binary MXene-based nanocomposites, together with various synthesis and coupling techniques for several photocatalytic applications.

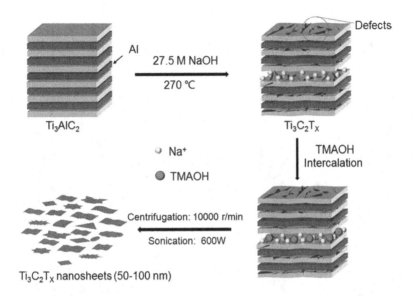

FIGURE 5.4 Synthesis of fluorine-free $Ti_3C_2T_x$ nanosheets via a two-step alkali route (Li et al. 2019).

Huang and his research team studied the influence of mass ratio of ultrathin MXene (Ti_3C_2) to indium sulfide (In_2S_3) (Huang et al. 2021). Ti_3C_2 nanosheets (MXene) can be obtained through the following steps: 0.8 g LiF was added to 10 mL HCl (9 mol L^{-1}) and stirred for 5 min. To prevent boiling, 0.5 g Ti_3AlC_2 was added slowly into the solution, which was then stirred for 24 h at room temperature for etching. The sample was centrifuged and washed using deionized water until a pH of 6.0 was reached. For delamination purposes, another 100 mL of deionized water was added, and the mixture shaken manually for 15 min. 2D transition-metal carbide Ti_3C_2 nanosheets (MXene) was obtained through centrifugation. The synthesis of Ti_3C_2 coupled with In_2S_3 was obtained through the following steps: 0.8 mmol $InCl_3 \cdot 4H_2O$ was added to 10 mL deionized water and stirred for 60 min. The desired amount of Ti_3C_2 was added and stirred for 60 min at room temperature. Then 1.2 mmol thioacetamide (TAA) was added and stirred for 15 min. The mixture was transferred to a 50 mL flask containing inert gas. The mixture was refluxed for 90 min at 95 °C. The mixture was then cooled down rapidly using an ice-water mixture and centrifugally washed with water and ethanol. The sample was dried overnight in a vacuum-drying chamber at 35 °C. Using this procedure, Ti_3C_2/In_2S_3 composite with mass ratio of Ti_3C_2 to In_2S_3 from 1% to 3% was obtained. The performance of their synthesized MXene composites was tested by photoreduction of Cr^{6+} under visible-light irradiation. The result of 6 min irradiation showed that the Ti_3C_2 coupled with In_2S_3 significantly enhanced the photoactivity when compared to those of bare Ti_3C_2 and In_2S_3. The photoactivities increased when the mass ratio of Ti_3C_2 to In_2S_3 increased from 1% to 2% whereas it started to drop when the mass ratio of Ti_3C_2 to In_2S_3 beyond 3%.

TABLE 5.1

Recent Development on the Preparation of Binary MXene-Based Nanocomposites for Various Photocatalytic Applications

| MXene Composition | | Synthesis Methods | Solvent or Surfactant | Reaction Condition | | Morphology | Pollutant Type and Concentration (mg L^{-1}) | Catalyst Loading (g L^{-1}) | Performance | Ref. |
MAX Phase	MXene Type			Temp. (°C)	Time (h)					
Ti$_3$AlC$_2$	Ti$_3$C$_2$/In$_2$S$_3$	Electrostatic self-assembly	1.2 mmol thioacetamide (TAA)	95	1.5	Nanosheet with nanoflakes	Cr^{6+}: 15	0.5	~100% reduction within 6 min	(Huang et al. 2021)
Ti$_3$AlC$_2$	Ti$_3$C$_2$/NiAl-LDH	Hydrothermal	100 ml Dimethyl sulfoxide (DMSO)	120	24	Flower like hierarchical	CO$_2$ reduction; Highly purified CO$_2$ was purged	250	CO production rate up to 11.82 μmol g^{-1}h^{-1} with 92% selectivity	(Shi et al. 2021)
Ti$_3$AlC$_2$	Ti$_3$C$_2$/Cd$_{0.5}$Zn$_{0.5}$S	Hydrothermal	25 mmol thioacetamide (TAA) + 10 mL ethylenediamine	160	24	Nanosheets decorated with Rod	H$_2$ production; DI water; seawater	0.5	H$_2$ production rate is up to 14.07 mmol h^{-1} g^{-1}	(Zeng et al. 2021)
Ti$_3$AlC$_2$	Ti$_3$AlC$_2$/MoS$_2$	Hydrothermal	DI water	210	24	Nanosheets	Ranitidine: 10	1	88.4% degradation within 1 h	(Zou et al. 2021)
Nb$_2$AlC	Nb$_2$CT$_x$/Bi$_2$WO$_6$	Hydrothermal	0.05 g cetrimonium bromide (CTAB)	120	24	Nanosheets	i. Rhodamine B (RhB): 15 ii. Methylene blue (MB): 15 iii. Tetracycline Hydrochloride (TC-HCl): 15	0.5	i. RhB: 92.7% degradation within 90 min ii. MB: 92.7% degradation within 90 min iii. TC-HCl: 83.1% degradation within 120 min	(Cui et al. 2020)
Nb$_2$AlC	Nb$_2$CT$_x$/Nb$_2$O$_5$	Hydrothermal	DI water	200	12–30	Nanosheets decorated with nanorods	i. RhB: 10 ii. TC-HCl: 10	1	i. RhB: 98.5% degradation within 120 min. ii. TC-HCl: 91.2% degradation within 180 min	(Cui et al. 2021)

(Continued)

TABLE 5.1 (Continued)
Recent Development on the Preparation of Binary MXene-Based Nanocomposites for Various Photocatalytic Applications

MAX Phase	MXene Type	Synthesis Methods	Solvent or Surfactant	Temp. (°C)	Time (h)	Morphology	Pollutant Type and Concentration (mg L^{-1})	Catalyst Loading (g L^{-1})	Performance	Ref.
Ti_3AlC_2	$Ti_3C_2T_x$/ZnO	Calcination	DI water	550	4	Nanosheets	i. Methyl orange (MO): 0.4 ii. RhB: 0.4	0.6	i. MO: 99.7% degradation within 50 min. ii. RhB: 99.8% degradation within 70 min	(Ta, Tran, and Noh 2020)
Ti_3AlC_2	$Ti_3C_2T_x$/ g-C_3N_4	Heating under nitrogen	DI water	250	2	Mesoporous	CO_2 reduction; Ar gas was purged for 5 min	50	CO production rate up to 3.98 $\mu mol\ g^{-1}h^{-1}$	(Li, Bai, et al. 2021)
Ti_3AlC_2	$Ti_3C_2T_x$/Cu_2O	Hydrothermal	DI water	90	5	Ultrathin sheet with hexapod	CO_2 reduction; CO_2 gas (< 1 atm) was purged	30 mg on quartz plate	CO production rate up to 17.55 $\mu mol\ g^{-1}h^{-1}$	(Zhang et al. 2021)
Ti_3AlC_2	$Ti_3C_2T_x$/ $SnNb_2O_6$	Hydrothermal	60 mL Dimethyl sulfoxide (DMSO)	85	8	Nanosheet	i. TC-HCl: 10 ii. RhB: 10 iii. H_2 production, 20% methanol	0.25 (photo-degradation) and 0.2 (H_2 evo-lution)	i. TCH: 70% degradation within 60 min. ii. RhB: 98% degradation within 60 min iii. H_2 production rate up to 175.2 $\mu molh^{-1}g^{-1}$	(Wang, Chen, et al. 2021)
Ti_3AlC_2	$Ti_3C_2T_x$/TiO_2	Hydrothermal oxidation	200 g Sodium Borohydrid-e ($NaBH_4$)	180	10	Layer structure with arrays	MO: 30	0.5	99% MO degraded within 40 min	(Hieu et al. 2021)

Recently, NiAl-LDH was coupled with Ti_3C_2 (Shi et al. 2021). To prepare Ti_3C_2, the Al atoms from Ti_3AlC_2 were selectively etched in a 40% HF solution. The resulting Ti_3C_2 was rinsed several times with deionized water and vacuum dried. The Ti_3C_2/NiAl-LDH composite was fabricated through the following steps: 1 g of Ti_3C_2 was dispersed in 100 mL of DMSO and stirred overnight. The sample was separated and sonicated in deionized water, followed by centrifugation. The supernatant was collected, and the concentration of Ti_3C_2 was analysed to be 15 mg mL^{-1}. The Ti_3C_2 dispersion was mixed with 0.87 g $Ni(NO_3)_2 \cdot 6H_2O$, 0.375 g $Al(NO_3)_3 \cdot 9H_2O$, 0.592 g NH_4F and 2.4 g urea. The mixture was sonicated and stirred for 30 min. The resulting solution was transferred into a 100 mL Teflon-lined container and autoclaved at 120°C for 24 h. The as-made sample was washed with deionized water and ethanol several times. The samples were labelled according to the volume of Ti_3C_2 dispersion (10 mL of Ti_3C_2 dispersion was denoted as LDH/TC−1, 30 mL of Ti_3C_2 dispersion was denoted as LDH/TC−2 while 60 mL of Ti_3C_2 dispersion was denoted as LDH/TC−3). The synthesis of NiAl-LDH was similarly conducted without the modification by Ti_3C_2 dispersion. The photocatalytic performances of the prepared samples were evaluated by CO_2 conversion under 300 W Xenon arc lamp irradiation. From their CO_2 photoconversion results, it was obvious that the production of CO increased after coupling with Ti_3C_2. Under xenon light irradiation, an enhancement of CO production under all the Ti_3C_2/NiAl-LDH composite was noticed. The highest CO production was obtained by LDH/TC−2 composite (11.82 $\mu mol\ g^{-1}h^{-1}$), which was 3.2 times higher than the bare NiAl-LDH. In addition, all the Ti_3C_2/NiAl-LDH composites exhibited impressive selectivity that was higher than 92%.

Zeng et al. (Zeng et al. 2021) reported on preparing the Ti_3C_2/$Cd_{0.5}Zn_{0.5}S$ composites by a hydrothermal method. The Ti_3C_2 MXene was prepared through the following procedures: 1 g of Ti_3AlC_2 and 100 mL of (40 wt%) HF acid were added into a 500 mL Teflon cup and stirred for 72 h at room temperature. Subsequently, the solution was centrifuged and the obtained sample was rinsed with distilled water several times until the pH became neutral. The resulting product (Ti_3C_2) was freeze-dried for 48 h. The $Cd_{0.5}Zn_{0.5}S$ was obtained through the following procedures: 20 mmol $Zn(Ac)_2 \cdot H_2O$ and 20 mmol $Cd(Ac)_2 \cdot 4H_2O$ were dissolved in 30 mL deionized water and stirred for 30 min. At the same time, 25 mmol thioacetamide, 10 mL ethylenediamine, and 20 mL deionized water were added into the mixture. The mixture was transferred to a 100 mL Teflon beaker and autoclaved for 12 h at 180 °C. The resulting samples ($Cd_{0.5}Zn_{0.5}S$) were rinsed with deionized water and freeze-dried overnight. Ti_3C_2/$Cd_{0.5}Zn_{0.5}S$ composites were prepared through the following routes: 0.1 g of $Cd_{0.5}Zn_{0.5}S$ was dissolved in 20 mL deionized water. Various amounts (1%, 3%, 5%, 7%) of Ti_3C_2 were dissolved in 10 mL deionized water and mixed with the previous solution for 2 h under argon gas condition. The mixture was transferred to a Teflon cup and autoclaved for 24 h at 160 °C. Subsequently, the resulting precipitate was centrifuged, washed with deionized water, and dried overnight in an oven at 60 °C. Their prepared catalysts were used for hydrogen evolution under visible light irradiation. The results revealed that the loading of Ti_3C_2 on $Cd_{0.5}Zn_{0.5}S$ enhanced the hydrogen-evolution performance. A higher photocatalytic hydrogen-production rate of 14.04 $mmolh^{-1}g^{-1}$ was obtained using 5% Ti_3C_2/$Cd_{0.5}Zn_{0.5}S$, which was five times higher than bare $Cd_{0.5}Zn_{0.5}S$.

However, a reduction in photocatalytic performance was found when the loading of Ti_3C_2 exceeded 5%, which was caused by the reduction of active sites due to excess loading of Ti_3C_2.

Zou et al. (Zou et al. 2021) prepared Ti_3C_2 from etching of Ti_3AlC_2 and Ti_3C_2/MoS_2 composites via a hydrothermal method. The detailed preparation method was as follows: 2 g Ti_3AlC_2 was added to 40 mL HF solution (49%) and stirred for 48 h to etch the Al from $TiAlC_2$. The solution was centrifuged at 8000 rpm, and the obtained product was rinsed with ultrapure water until the pH was more than 5. The resulting product (Ti_3C_2) was dried overnight in an oven at 60°C. The conventional hydrothermal method was used in the synthesis of Ti_3C_2/MoS_2 composite as follows: Ti_3C_2 powder was dispersed in distilled water to form 4 mg mL^{-1} Ti_3C_2 suspension. The suspension was diluted to 1 mg mL^{-1} in a water-ethanol solution with (water:ethanol = 10:1) and sonicated for 3 h for exfoliation. Then 242 mg $Na_2MoO_4·2H_2O$ and 380 mg NH_2CSNH_2 were mixed with the previous suspension and autoclaved for 24 h at 210°C. The samples were labelled according to the amount of Ti_3C_2 added, 1.21 mL of 4 mg mL^{-1} denoted as MT−1, 4.84 mL denoted as MT−2, 9.68 mL denoted as MT−3, 24.2 mL denoted as MT−4, and 36.3 mL denoted as MT−5. Figure 5.5 shows the main preparation steps on the synthesis of Ti_3C_2 and Ti_3C_2/MoS_2 composites. Pure MoS_2 was fabricated using the same procedures without adding Ti_3C_2. The effectiveness of their prepared photocatalysts was tested on the photodegradation of ranitidine and reduction of nitrosamine dimethylamine under visible light irradiation. The results showed that the coupling of Ti_3C_2 with MoS_2 greatly enhanced the photocatalytic performance with the highest degradation efficiency of 88.4% achieved by MT−4 within 60 min irradiation. MT−5 showed slightly inferior results in photodegradation performance due to the excessive amount of Ti_3C_2 added.

Cui et al. (Cui et al. 2020) prepared a Nb_2CT_x/Bi_2WO_6 composite and used it in the photodegradation of RhB and MB dyes under visible-light irradiation. They

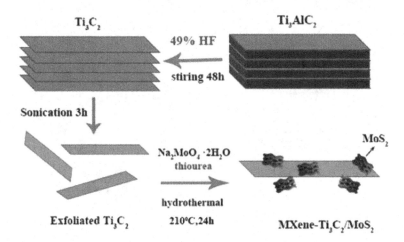

FIGURE 5.5 Synthesis routes in preparation of Ti_3C_2 sheets and Ti_3C_2/MoS_2 composites (Zou et al. 2021).

dispersed 1 g of Nb_2AlC in 60 mL HF solution ($\geq 40\%$) and stirred it for 90 h at 55°C. The resulting products were washed with deionized water until the pH reached 7.0. The obtained powder was re-dispersed in 10 mL 26% tetra-methylammonium hydroxide (TMAOH) for three days. The suspension was cen-trifuged and washed in deionized water to remove TMAOH. The sample was again dispersed in deionized water, sonicated for 1 h, and finally centrifuged to obtain the Nb_2CT_x. The preparation method of Nb_2CT_x/Bi_2WO_6 composite was as follows: 0.5 g of CTAB and 1 mmol of $Na_2WO_4 \cdot 2H_2O$ were dissolved in 80 mL deionized water and stirred until a clear solution resulted. Different mass ratios of Nb_2CT_x and $(NO_3)_3 \cdot 5H_2O$ were added to the above solution and stirred for 1 h. The solution was then transferred to a Teflon-lined beaker and heated for 24 h at 120°C. The obtained samples were rinsed with ethanol and water and dried in an oven at 60°C. The samples with 0.5, 2.0, 5.0 and 10.0 wt% of Nb_2CT_x were denoted as BN–0.5, BN–2.0, BN–5.0 and BN–10.0, respectively. The synthetic route of Nb_2CT_x/Bi_2WO_6 composites from Nb_2AlC (MAX) is illustrated in Figure 5.6. The pseudo-first-order kinetic rate constant, evaluated within 90 min reaction, showed that all the Nb_2CT_x/Bi_2WO_6 composites had higher degradation rates for the photo-degradation of RhB and MB under xenon lamp irradiation compared to the bare Bi_2WO_6. BN–2 exhibited a constant rate of (k) of 0.072 min^{-1}, which was 2.8 times higher than the pristine Bi_2WO_6 in RhB degradation.

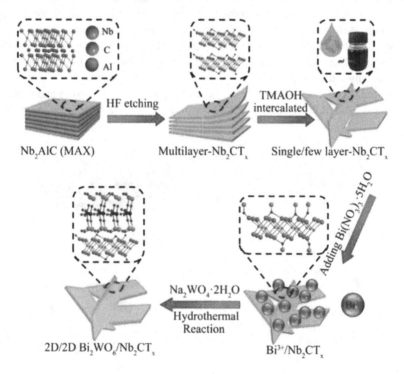

FIGURE 5.6 Schematic diagram of Nb_2CT_x/Bi_2WO_6 composites synthesis process (Cui et al. 2020).

In 2021, Cui et al. (Cui et al. 2021) prepared a rod-like Nb_2O_5/Nb_2CT_x composite via a hydrothermal method. As discussed before, Nb_2CT_x was prepared through the following steps: 1 g of Nb_2AlC dispersed into 60 mL ($\geq 40\%$) HF solution and stirred for 90 h at 55°C. The resulting product was centrifuged, rinsed with deionized water several times until the pH was more than 6.0, and dried in an oven to obtain Nb_2CT_x. The detailed synthesis procedure for Nb_2O_5/Nb_2CT_x composite was as follows: 60 mg of Nb_2CT_x was dispersed in 150 mL deionized water and stirred for 1 h to form suspension. The suspension was transferred to a Teflon-lined cup and underwent hydrothermal oxidation at 200°C for 12, 18, and 30 h. The obtained Nb_2O_5/Nb_2CT_x composites were named according to their hydrothermal duration: Nb_2CT_x–12, Nb_2CT_x–18, and Nb_2CT_x–30, respectively. Figure 5.7 illustrates the synthetic procedures of Nb_2O_5/Nb_2CT_x composites. Photodegradation of RhB and TC-HCl under visible light were considered to test the performances of their prepared catalysts. Their results showed that the photoactivity of Nb_2O_5/Nb_2CT_x composites in 120 min degradation of RhB increased from about 24.9% (pure Nb_2O_5) to 98.5% (Nb_2CT_x–30). In addition, the results showed that the photodegradation of TC-HCl over the Nb_2CT_x–30 was 91.2%, and its degradation rate was 13.5 times and 6.8 times higher than Nb_2CT_x and pure Nb_2O_5, respectively. The Nb_2CT_x–30 exhibited superior performance, owing to the formation of a heterojunction between the Nb_2O_5 and Nb_2CT_x, resulting in rapid transfer and separation of photogenerated carriers.

Apart from the above-described methods of preparing hybrid MXene, Ta et al. (Ta, Tran, and Noh 2020) reported the utilization of calcination method to modify $Ti_3C_2T_x$ MXene with ZnO. The $Ti_3C_2T_x$ was selectively etching Al layers from the Ti_3AlC_2 MAX phase. Then 1 g of Ti_3AlC_2 was dispersed and stirred in 28 mL HF solution. The dispersion was further stirred in an oil bath at 50°C for 24 h. The precipitate was rinsed with deionized water until the pH reached 7.0; it was then dried overnight in an oven at 60°C. The $Ti_3C_2T_x$/ZnO composites were prepared

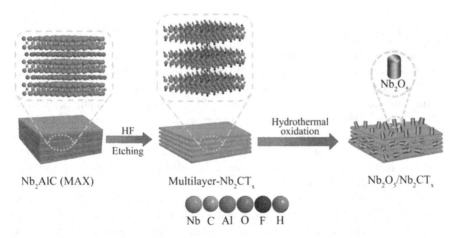

FIGURE 5.7 Schematic diagram of the preparation of rod-like Nb_2O_5/Nb_2CT_x composites (Cui et al. 2021).

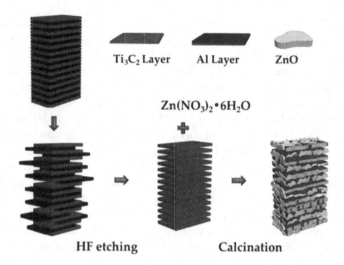

FIGURE 5.8 Schematic diagram of synthetic process for ZnO/Ti$_3$C$_2$T$_x$ composites (Ta, Tran, and Noh 2020).

through the following steps: various amounts of Ti$_3$C$_2$T$_x$ (12, 20, 30 mg) were dispersed in 5 mL of Zn^{2+} (1.34 M) solution and dried in an oven. The resulting product was transferred into a crucible and calcined at 550°C for 4 h. The obtained samples were named as ZnO/Ti$_3$C$_2$T$_x$–1 (10 mg), ZnO/Ti$_3$C$_2$T$_x$–2 (20 mg), and ZnO/Ti$_3$C$_2$T$_x$–3 (30 mg), respectively. The synthetic route of ZnO/Ti$_3$C$_2$T$_x$ composites is illustrated in Figure 5.8. The photocatalytic performance of ZnO/Ti$_3$C$_2$T$_x$ composites was tested with photodegradation of MO and RhB lamp under Xenon lamp. Their synthesized ZnO/Ti$_3$C$_2$T$_x$ composites revealed excellent photocatalytic performance for MO and RhB, three times faster than the pure ZnO under simulated solar-light irradiation.

5.2.3 Preparation of Ternary MXene-Based Nanocomposites

The coupling of the accordion-like or the few/single layer 2D MXene nanosheets with other semiconductor photocatalysts often involved techniques such as electrostatic self-assembly, electrophoretic coating, drop casting, spin-coating, spray-coating, and doctor-blade methods. A ternary MXene-based hybrid photocatalyst is defined as the formation of complex MXene-semiconductor semiconductor configuration, which the ternary system comprised of three distinctive component catalysts. Table 5.2 summarizes the recent development of ternary MXene-based nanocomposites together with various synthesis and coupling techniques for several photocatalytic applications.

Among the available methods, ultrasonic or mechanical stirring/mixing assisted self-assembly techniques are the most widely adopted method by researchers to synthesize the ternary MXene-based composites. α-Fe$_2$O$_3$/ZnFe$_2$O$_4$@Ti$_3$C$_2$ nanocomposites with different loadings of α-Fe$_2$O$_3$ (5, 10 and 15 wt%) were prepared through an ultrasonic self-assembly route to eliminate Rhodamine B (RhB) and Cr(VI)

TABLE 5.2

Recent Development on the Preparation of Ternary MXene-Based Nanocomposites for Various Photocatalytic Applications

MAX Phase	MXene Type	Synthesis Methods	Solvent or Surfactant	Temp. (°C)	Time (h)	Morphology	Pollutant Type and Concentration (mg L^{-1})	Catalyst Loading (g L^{-1})	Performance	Ref.
Ti$_3$AlC$_2$	Ti$_3$C$_2$/TiO$_2$/ g-C$_3$N$_4$ (CN/TCT)	Ti$_3$C$_2$/TiO$_2$ (TCT): In-situ growth oxidation	HF ~ 39–49 wt%	Room condition	24–96	Packed layer by layer structure with TiO$_2$ nucleated nanoparticles	CO$_2$/CH$_4$ = 2:1	0.1	Generation of CO: 73.2% within 4 h	(Khan and Tahir 2021)
		Ti$_3$C$_2$/TiO$_2$/g-C$_3$N$_4$: Ultrasonic self-assembly	Ethanol	Room condition	1	nanoparticles	CO$_2$/CH$_4$ = 1:2		Generation of H$_2$: 43% within 4 h	
Ti$_3$AlC$_2$	Ag/Ag$_3$PO$_4$/ 3 wt% Ti$_3$C$_2$	Electrostatic self-assembly	DI water and 0.1 mol L^{-1} methanol	Room condition	4	Spherical Ag$_3$PO$_4$ + accordion-like multilayer Ti$_3$C$_2$	i. MO: 20 ii. Cr(VI): 10	0.5	i. MO: 93% degradation within 1 h ii. Cr^{6+}: 61% reduction within 1 h	(Sun, Tao, et al. 2021)
Ti$_3$AlC$_2$	Ti$_3$C$_2$T$_x$/ (001)TiO$_2$/ C$_3$N$_4$	Protonation of g-C$_3$N$_4$	HCl – 0.5 mol L^{-1}	Room condition	5	Nanosheets within nanosheets	i. CO$_2$ reduction: High purity gas purged for 30 mins	50	Generation of CO: 30 ppmGeneration of CH$_4$: 6.34 ppm	(Wu et al. 2021)
		Ti$_3$C$_2$T$_x$/ (001)TiO$_2$/ C$_3$N$_4$: Facile in-situ oxidation with	DI water	Room condition	N/A		ii. H$_2$ generation: triethanolamine	0.5	Generation of H$_2$: ~143.2 µmol g^{-1} h^{-1}	

(Continued)

TABLE 5.2 (Continued)

Recent Development on the Preparation of Ternary MXene-Based Nanocomposites for Various Photocatalytic Applications

MAX Phase	MXene Type	Synthesis Methods	Solvent or Surfactant	Reaction Condition Temp. (°C)	Time (h)	Morphology	Pollutant Type and Concentration (mg L^{-1})	Catalyst Loading (g L^{-1})	Performance	Ref.
Ti$_3$AlC$_2$	BiVO$_4$@ZnIn$_2$S$_4$/Ti$_3$C$_2$ quantum dots	BiVO$_4$@ZnIn$_2$S$_4$: *In-situ* growth combined with two-step solvothermal; electrostatic self-assembly	DI water	95	2	Hierarchical core-shell flower-like microsphere	Water (TEOA, 15 vol%)	3	Generation of O$_2$: 50.83 µmol g^{-1} h^{-1} Generation of H$_2$: 102.67 µmol g^{-1} h^{-1}	(Du et al. 2020)
		BiVO$_4$@ZnIn$_2$S$_4$/Ti$_3$C$_2$: Ultrasonic and mechanical stirring	DI water	40	N/A		Bisphenol A: 10	1.75	Photodegradation on Bisphenol A: ~96.4% within 180 min	
Ti$_3$AlC$_2$	10 wt% α-Fe$_2$O$_3$/ZnFe$_2$O$_4$@Ti$_3$C$_2$	α-Fe$_2$O$_3$/ZnFe$_2$O$_4$: One-step hydrothermal; Fe$_2$O$_3$/ZnFe$_2$O$_4$@Ti$_3$C$_2$: Ultrasonic assisted self-assembly	Fe$_2$O$_3$/ZnFe$_2$O$_4$: DI water + EG; Ethanol + Nafion	180; Room condition	12; 30	Spindle-shaped α-Fe$_2$O$_3$ + lamellar nanolayers Ti$_3$C$_2$	i. RhB: 10 ii. Cr(VI): 10	0.2	i. RhB: complete degradation within 150 min ii. Cr^{6+}: 100% reduction to Cr^{3+} within 90 min	(Zhang et al. 2020)

(Continued)

TABLE 5.2 (Continued)

Recent Development on the Preparation of Ternary MXene-Based Nanocomposites for Various Photocatalytic Applications

MAX Phase	MXene Composition — MXene Type	Synthesis Methods	Solvent or Surfactant	Reaction Condition — Temp. (°C)	Reaction Condition — Time (h)	Morphology	Pollutant Type and Concentration (mg L^{-1})	Catalyst Loading (g L^{-1})	Performance	Ref.
Ti$_3$AlC$_2$	6% RuO$_2$@Ti-O$_2$–Ti$_3$C$_2$	Hydrothermal	N/A	150	10	Ti$_3$C$_2$ accordion-like structure with decorated larger TiO$_2$ and smaller RuO$_2$ nanoparticles	Pure water purged with pure N$_2$ gas (50 mL min^{-1})	0.5	N$_2$ reduction: 425 μmol l^{-1} g^{-1} within 450 min	(Hao et al. 2020)
Ti$_3$AlC$_2$	Ti$_3$C$_2$@Ti-O$_2$/ZnIn$_2$S$_4$-50 wt%	Two-step hydrothermal with *In-situ* oxidation	DI water + thioacetamide, TTA (0.56 mmol)	180	3	Layered mesoporous	Water + 0.35 M of Na$_2$S.9H$_2$O and 0.25 M of Na$_2$SO$_3$	0.1875	Generation of H$_2$: 1185.8 μmol g^{-1} h^{-1} within 4 h	(Huang, Li, and Meng 2020)
Nb$_2$AlC	CdS/Nb$_2$O$_5$/Nb$_2$C	One-pot *in-situ* hydrothermal	Ultrapure water + thioacetamide (TTA)	180	21	Nanorods-nanoparticles embedded inside the interlayer space of Nb$_2$C nanosheets	Carbamazepine (CBZ): 5	0.1	CBZ: 92% degradation within 4 h	(Zu et al. 2020)
Ti$_3$AlC$_2$	Ti$_3$C$_2$/TiO$_2$/Co-0.5%	Two-step calcination	N$_2$	400 (5 °C min^{-1})	1	Hierarchical accordion-like with rough surface	Pure water purged with pure N$_2$ gas	0.3	Production of NH$^+_4$: 110 μmol g^{-1} h^{-1}	(Gao et al. 2021)
Ti$_3$AlC$_2$	TiO$_2$-Ti$_3$C$_2$/Ru-20%	In-situ formation TiO$_2$ and direct reduction	DI water + NaBF$_4$ + 1 M HCl	180	10–36	Ti$_3$C$_2$ accordion-like structure decorated with TiO$_2$ and smaller Ru nanoparticles	10% ethanol solution	0.2	Generation of H$_2$: 235.3 μmol g^{-1} h^{-1}	(Liu, Li, et al. 2020)

under a 300 W Xe lamp [20]. The ternary α-Fe$_2$O$_3$/ZnFe$_2$O$_4$@Ti$_3$C$_2$ nanocomposites were then synthesized by ultrasonicating 18 mg Ti$_3$C$_2$ with 10 mL ethanol for 10 min, followed by dispersing 2 mg of fresh α-Fe$_2$O$_3$/ZnFe$_2$O$_4$ powder and 20 µl Nafion solution (5 wt%) into the solution. The resultant solution was continuously sonicated for 20 min to allow the deposition of the α-Fe$_2$O$_3$/ZnFe$_2$O$_4$ onto the lamellar layers of Ti$_3$C$_2$ (Figure 5.9a). They reported that the 10 wt% α-Fe$_2$O$_3$/ZnFe$_2$O$_4$@Ti$_3$C$_2$ had a complete photodegradation of RhB within 150 min and fully eliminated the Cr^{6+} after 90 min. The reason given for this result was the formation of intimate interfaces among the three catalytic materials in granting the high-separation efficiency of the electron-hole pair and electron transfer on the surface of the catalyst for the excellent photocatalysis (Figure 5.9b).

Khan and Tahir (Khan and Tahir 2021) prepared a densely packed multilayer ternary Ti$_3$C$_2$/TiO$_2$/g-C$_3$N$_4$ (TCT/CN) photocatalyst via an ultrasonic-assisted self-assembly route for the photoconversion of mixed CO$_2$/CH$_4$ gas under visible light. The TCT/CN composite was produced by adding 1.0 g of as-synthesized TCT and 1.0 g of g-C$_3$N$_4$ samples into two separated solutions containing 20 mL ethanol. Both solutions were stirred for 30 min at room temperature. The g-C$_3$N$_4$ solution was then transferred to the TCT solution and followed by ultrasonic treatment to mix the two suspensions for 1 h. The self-assembled TCT/CN sample were finally dried in an oven at 100°C for a day. The researchers had reported that the ternary

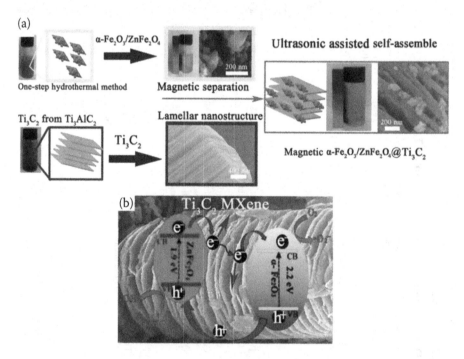

FIGURE 5.9 (a) Schematic elucidation on the synthesis of the ternary magnetic α-Fe$_2$O$_3$/ZnFe$_2$O$_4$@Ti$_3$C$_2$ and (b) the schematic mechanism of magnetic α-Fe$_2$O$_3$/ZnFe$_2$O$_4$@Ti$_3$C$_2$ MXene photocatalyst (Zhang et al. 2020).

TCT/CN composite gave the best photoconversion performance of mixed gas at 30.17 µmol and a maximum H_2 production rate at 19.55 µmol within 4 h. The photoconversion enhancement was attributed to the efficient charge separation on the formation of the heterojunction structures between Ti_3C_2-TiO_2 and g-C_3N_4. They added that the ternary TCT/CN photocatalyst structure can endow the high number of basic sites for the strong CO_2 adsorption process.

A series of Ag/Ag_3PO_4/Ti_3C_2 photocatalysts with different Ti_3C_2 loadings (1, 3, 5 wt%) was synthesized using the electrostatic self-assembly method and Ag^+ partial reduction in Ti_3C_2 for the photodegradation of MO and photoreduction of Cr^{6+} under visible-light irradiation (Sun, Tao, et al. 2021). The detail fabrication of Ag/Ag_3PO_4/Ti_3C_2 photocatalyst was as follows: The as-produced Ti_3C_2 underwent intercalation using 12 mL dimethyl sulfoxide (DMSO) organic solvent and followed by an ultrasonication to produce single or few-layers Ti_3C_2 nanosheets. After that, the Ti_3C_2 nanosheets colloid (0.82 mg mL^{-1}) in deionized water was prepared. The Ti_3C_2 nanosheets colloid dispersion and 0.6087 g $AgNO_3$ were placed into deionized water under stirring to form 40 mL of solution A. Meanwhile, 0.4278 g $Na_2HPO_4 \cdot 12H_2O$ was dissolved in 40 mL deionized water and stirred for 30 min to form solution B. The solutions A and B were then stirred for 4 h at room condition to form the ternary Ag/Ag_3PO_4/Ti_3C_2 photocatalyst. They found that the optimized Ag/Ag_3PO_4/3 wt% Ti_3C_2 sample can achieve the remarkable MO photodegradation and Cr^{6+} photoreduction with efficiencies of 93% and 61%, respectively.

An in situ oxidation treatment combined with an electrostatic self-assembly route was used to produce the $Ti_3C_2T_x$/(001)TiO_2/C_3N_4 nanosheets (NSs) and applied on the photoreduction of CO_2 and photogeneration of H_2 under sunlight exposure (Wu et al. 2021). Initially, 0.33 g $NaBF_4$ was added into 13.75 mL Ti_3C_2 solution (~5 mg mL^{-1}) and followed by the addition of 1.25 mL HCl into the solution. The suspension was then transferred to a Teflon-lined stainless-steel autoclave for hydrothermal treatment at 180 °C for 4 h to form binary $Ti_3C_2T_x$/(001)TiO_2 NSs. Subsequently, 100 mg of as-synthesized g-C_3N_4 nanosheets was added into 20 mL HCl under ultrasonication for 4 h to yield the protonated g-C_3N_4 NSs solution. The ternary photocatalyst was synthesized by dispersing 7 mg $Ti_3C_2T_x$/(001)TiO_2 NSs in 7 mL deionized water to get solution A. Afterward, 3 mL of protonated g-C_3N_4 NSs solution was added into solution A under vigorous stirring. The resultant $Ti_3C_2T_x$/(001)TiO_2/C_3N_4 NSs was dried at 50 °C under vacuum for 12 h. Their findings revealed that the $Ti_3C_2T_x$/(001)TiO_2/C_3N_4 NSs achieved a four-fold enhancement on the photocatalytic H_2 generation and a three-fold improvement on the photocatalytic CO_2 reduction as compared to the pure (001)TiO_2 or g-C_3N_4 NSs.

Du et al. (Du et al. 2020) synthesized a novel all-solid-state hierarchical core-shell flower-like microsphere direct Z-scheme $BiVO_4$@$ZnIn_2S_4$/Ti_3C_2 quantum dots (QDs) photocatalyst through an in situ growth two-step hydrothermal together with ultrasonic self-assembly method. The Ti_3C_2 QDs was first prepared by adding 0.5 g Ti_3C_2 into 60 mL deionized water in N_2 atmosphere under ultrasonication for 10 min. Thereafter, the pH of the solution was adjusted to 6.0 using ammonia solution, and the mixture was hydrothermally treated at 100 °C for 6 h in a 100 mL Teflon-lined autoclave. To fabricate the ternary $BiVO_4$@$ZnIn_2S_4$/Ti_3C_2, 0.03 g of Ti_3C_2 QDs was dispersed into 30 mL deionized water under ultrasonication for 15 min. This step was followed by dispersing 0.3 g BiVO$_4$@ZnIn$_2$S$_4$ on the Ti$_3$C$_2$ QDs suspension under vigorous stirring at 40 °C to obtain the ultrasonic-assisted self-assembly attached Ti$_3$C$_2$ QDs onto the BiVO$_4$@ZnIn$_2$S$_4$. The ternary BiVO$_4$@ZnIn$_2$S$_4$/Ti$_3$C$_2$ QDs exhibited superior water-splitting into O$_2$ and H$_2$ evolution rates up to 50.83 and 102.67 μmol g^{-1} h^{-1} (~1:2), respectively. The results of their investigations revealed the formation of a Schottky barrier between the co-catalyst Ti$_3$C$_2$ QDs at the interface in contact with ZnIn$_2$S$_4$ and the excellent metal conductivity of Ti$_3$C$_2$ QDs had further promoted the charge separation of the photocatalysts.

Using a hydrothermal method, Liu et al. (Liu, Li, et al. 2020) synthesized the ternary TiO$_2$-Ti$_3$C$_2$/Ru photocatalysts and employed them for H$_2$ evolution under visible-light irradiation. The TiO$_2$-Ti$_3$C$_2$/Ru synthesis process was as follows: layered Ti$_3$C$_2$/Ru was initially prepared via ultrasonicating 200 mg Ti$_3$C$_2$ into 200 mL deionized water under N$_2$ atmosphere for 30 min. Afterward, RuCl$_3$·xH$_2$O (Ti$_3$C$_2$/Ru^{3+} = 100:2) was added to form a uniform suspension and was vigorously stirred for 6 h under N$_2$ condition. Next 80 mg of as-prepared Ti$_3$C$_2$/Ru and 0.165 g NaBF$_4$ was dispersed into 15 mL of 1 M HCl aqueous solution. The solution was then ultrasonicated for 20 min and stirred for 10 min. The resultant suspension solution was hydrothermally treated at 180 °C to form the ternary TiO$_2$-Ti$_3$C$_2$/Ru (Figure 5.10). The best TiO$_2$-Ti$_3$C$_2$/Ru–20 sample exhibited higher photocatalytic H$_2$ generation performance with 235.3 μmol g^{-1} h^{-1} than the Ru-TiO$_2$ (53 μmol g^{-1} h^{-1}). They claimed that the distinct design of the TiO$_2$-Ti$_3$C$_2$/Ru can concentrate the Ru onto a single spot on the Ti$_3$C$_2$. In this case, the TiO$_2$ can efficiently restrain the recombination of the electron-hole pairs and therefore improved the photocatalytic performance.

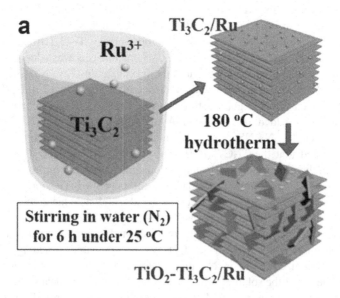

FIGURE 5.10 Schematic diagram on the preparation of TiO$_2$-Ti$_3$C$_2$/Ru using a hydrothermal technique (Liu, Li, et al. 2020).

A series of $Ti_3C_2@TiO_2/ZnIn_2S_4$ photocatalysts with different $ZnIn_2S_4$ loadings (30, 40, 50, and 60 wt%) were produced via a two-step hydrothermal and used for the hydrogen production under visible-light irradiation (Huang, Li, and Meng 2020). The $Ti_3C_2@TiO_2$ was first produced by adding 100 mg of freshly etched Ti_3C_2 and 165 mg of $NaBF_4$ into 1 M of HCl (15 mL) under stirring for 30 min. The solution was then transferred into a 25 mL autoclave and treated at 160 °C for 12 h. This step was followed by another hydrothermal treatment step to synthesize the $Ti_3C_2@TiO_2/ZnIn_2S_4$ photocatalyst. Typically, 200 mg of $Ti_3C_2@TiO_2$, 0.14 mmol of $ZnCl_2$ and 0.28 mmol $InCl_3$ were added in 65 mL deionized water under stirring and followed by the ultrasonic treatment for 1 h. The obtained solution was then added with 0.56 mmol TAA under continuous stirring for 30 min. The mixed suspension was poured into a 100 mL well-concealed Teflon-lined stainless-steel autoclave and heated at 180 °C for 3 h to form the ternary $Ti_3C_2@TiO_2/ZnIn_2S_4$ photocatalyst. The enhancement reason of photocatalytic H_2 evolution was originated to the remarkable light harnessing of $ZnIn_2S_4$ and Ti_3C_2, adequate active sites of Ti_3C_2, closed interfacial contact, and effective charge-carrier separation and transfer through the heterojunction structure.

A novel $CdS/Nb_2O_5/Nb_2C$ photocatalyst was fabricated through a facile one-pot in situ hydrothermal route and utilized for the photodegradation of carbamazepine (CBZ) under visible-light irradiation (Zu et al. 2020). The $CdS/Nb_2O_5/Nb_2C$ photocatalyst was obtained by first producing the Nb_2C MXene via etching for 168 h under room temperature from the 1.0 g Nb_2AlC parent precursor using 50 mL HF solution (40 wt%). Subsequently, the Nb_2O_5/Nb_2C was prepared by adding 200 mg Nb_2C into 40 mL ultrapure water under 30 min of ultrasonification and heated at 180 °C for 21 h in a 100 mL Teflon-lined autoclave. The freshly synthesized Nb_2O_5/Nb_2C, together with 100 mg Nb_2C MXene, were then added into 40 mL ultrapure water for a hydrothermal treatment at 180 °C for 21 h to yield the ternary $CdS/Nb_2O_5/Nb_2C$ photocatalyst. The $CdS/Nb_2O_5/Nb_2C$ sample was able to photodegrade 92% of CBZ within 4 h in comparison with the pure CdS (75% degradation).

Liu et al. (Liu, Yang, Jin, et al. 2020) exploited an one pot in situ growth methodology to produce the ternary $Ti_3C_2/TiO_2/BiOCl$. The typical synthesis was primarily involved by adding 0.0115 mol $Bi(NO_3)_3.5H_2O$ into a solution containing the desired amount of Ti_3C_2 and acetic acid (40 mL). The solution was then ultrasonicated and labeled as solution A. Another solution consisted of 0.0115 mol KCl with 20 mL distilled water dropped into solution A under vigorous agitation for 1 h. The resultant solution was poured slowly into a 100 mL Teflon-lined stainless-steel autoclave and hydrothermal for 24 h at 180 °C to create the ternary $Ti_3C_2/TiO_2/BiOCl$. The obtained sample was ultimately collected by filtration and washing steps and followed by drying at 60 °C for a day. The synthesized ternary samples displayed excellent photodegradation of RhB and tetracycline under sunlight irradiation, as compared to those of the binary composites (Ti_3C_2/TiO_2, $Ti_3C_2/BiOCl$, $TiO_2/BiOCl$) and pristine (TiO_2 and BiOCl) samples.

The study of Hao et al. (Hao et al. 2020) used a one-pot hydrothermal approach to prepare the $RuO_2@TiO_2-Ti_3C_2$ toward N_2 fixation under a Xe lamp irradiation (100 mW/cm^2). The Ti_3C_2 MXene was initially prepared by immersing into the MAX precursors in HF solution (40%) under constant stirring for two days. The

$RuO_2@TiO_2–Ti_3C_2$ composite was then fabricated by a hydrothermal treatment using $RuCl_3·3H_2O$ (50 mg) and HF etched Ti_3C_2 MXene (500 mg) treated at 150 °C for 10 h. Moreover, different loadings of RuO_2 (0%, 3%, 6%, and 9%) on $TiO_2–Ti_3C_2$ were also prepared by researchers using the similar procedures. Their optimum loading level seem to be 6% $RuO_2@TiO_2–Ti_3C_2$ on the N_2 reduction performance and reached 425 μmol l^{-1} g^{-1} after 450 min. The different contents of RuO_2 in the prepared photocatalysts also exerted different effects on the N_2 fixation. The reason they gave for the improvement was that the presence of ultra-small RuO_2 nanoparticles (2–4 nm) with a significant amount of metallic Ru(0) centers can act as effective N_2 activation sites.

A novel two-step calcination strategy was used to produce the ternary $Ti_3C_2/TiO_2/$ Co for N_2 fixation under UV-vis irradiation (Gao et al. 2021). The Ti_3C_2 was initially calcined at 400 °C under the N_2 gas (100 mL min^{-1}) for 1 h with ramping rate set at 5 °C min^{-1}. Thereafter, the calcined Ti_3C_2 was added into a solvent ($Co(NO_3)_2·6H_2O$ and 3 mL ethanol) and underwent ultrasonication for 60 min. The resulting sample was dried at 60 °C for 12 h and followed by the similar calcination in the first stage of calcination. The obtained powder was finally collected and labeled as $Ti_3C_2/TiO_2/$ Co–0.2%, $Ti_3C_2/TiO_2/Co–0.5\%$ and $Ti_3C_2/TiO_2/Co–1\%$ with respective to the mass ratios of Ti_3C_2 and $Co(NO_3)_2.6H_2O$ set as 500:1, 200:1 and 100:1, respectively. They also found that the $Ti_3C_2/TiO_2/Co–0.5\%$ displayed the best NH_4^+ production for nitrogen fixation at a rate of 110 μmol g^{-1} h^{-1} as compared to pristine materials (Ti_3C_2 and TiO_2), binary materials (Ti_3C_2/TiO_2, $TiO_2/Co–0.5\%$ and $Ti_3C_2/Co–0.5\%$), and other loadings of Co. They explained that the maximum active reactive sites induced by the optimal loading of the Co metal had modulated the binding sites of the catalyst to react with the reactants or products.

In summary, various methods, such as one-pot hydrothermal, two-step calcination, and in situ growth, were successfully used to prepare the ternary MXene-based nanocomposite materials for various photocatalytic applications. The ternary MXene-based photocatalyst demonstrated superior photocatalytic performance as compared to those of pristine system and heterojunction binary materials. This improvement was due to the successful formation of the ternary structure accompanied with the synergistic interface between the three distinctive materials, which usually leads to more active sites and the restraint of electron-hole pairs under light irradiation.

5.2.4 Preparation of Supported-MXene Composites

The MXene-supported composite had aroused tremendous interest of numerous researchers due to its excellent properties. In recent years, the MXene-supported composite was reported to be synthesized through various methods, including vacuum-assisted filtration, solution film casting via membrane coater, the electrospinning technique, the pressure-extrusion process, and simple magnetic mixing (Feng et al. 2020; Weng et al. 2020; Yang et al. 2020; Qian et al. 2021; Wang, Shao, et al. 2021). The unique and special traits of MXene allowed it to be fabricated as the supported-MXene composites (Jaya Prakash and Kandasubramanian 2021). Table 5.3 illustrates numerous MXene-supported composite materials prepared at different methods.

TABLE 5.3

Recent Findings of Various Synthesis Methods of MXene-Supported Composites

MXene Composition		Synthesis Methods	Solvent or Surfactant	Reaction Condition		Morphology	Pollutant Type and Concentration (mg L^{-1})	MXene Loading (wt%)	Performance	Ref.
MAX Phase	MXene Type			Temp. (°C)	Time (h)					
Ti$_3$AlC$_2$	PVA/SSA/Ti$_3$C$_2$ MXene/PTFE	Phase inversion film casting	NA	90	–	Nanosheet	NaCl; 0.6 M	2	99.8% salt rejection within 3 H	(Yang et al. 2020)
Ti$_3$AlC$_2$	Ti$_3$C$_2$ MXene/CA	Phase inversion film casting	1:1 Acetone/Acetic acid	25	24	Nanosheet	Treated sewage effluent	4	17.5% removal of total dissolved solids in 1400 mins	(Alfahel et al. 2020)
Ti$_3$AlC$_2$	Ti$_3$C$_2$@CA membrane	Film casting followed by formaldehyde crosslink	1:1 Acetone/Acetic acid	25	24	Nanosheet	Na$_2$SO$_4$; 2000	10	58.6% Na$_2$SO$_4$ rejection within 60 mins	(Pandey et al. 2020)
Ti$_3$AlC$_2$	Cellulose microsphere@void@Ti$_3$C$_2$ MXene	Pressure-extrusion process	NA	R.T.	5	Nanosheet	NA	CM: MXene = 1:7	EMI shielding at 59 dB	(Qian et al. 2021)
Ti$_3$AlC$_2$	Ti$_3$C$_2$ MXene/natural rubber	Pressure-extrusion process	Dicumyl peroxide	45–50	–	Nanosheet	NA	MXene: natural rubber = 10:1	Tensile strength at 25.5 MPa	(Liu et al. 2018)

(Continued)

TABLE 5.3 (Continued)
Recent Findings of Various Synthesis Methods of MXene-Supported Composites

MAX Phase	MXene Composition (MXene Type)	Synthesis Methods	Solvent or Surfactant	Reaction Condition Temp. (°C)	Time (h)	Morphology	Pollutant Type and Concentration (mg L^{-1})	MXene Loading (wt%)	Performance	Ref.
Ti_2AlC	Ti_2CT_x NS/PDDA	Simple chemical dropwise	PDDA	R.T. under Ar gas	2	Nanosheet	Re (VII); 100	0.4 g L^{-1}	q_e = 363 mg g^{-1}	(Wang et al. 2019)
Ti_3AlC_2	Alk-MXene/LDH	Simple mechanical stirring	NA	R.T.	24	Nanosheet	Ni^{2+}; 100	MXene: LDH = 1:1	Q_m = 222.7 mg g^{-1}	(Feng et al. 2020)
Ti_3AlC_2	ANF/ Ti_3C_2 MXene	Vacuum-assisted filtration	DMSO	R.T.	–	Nanosheet	NA	40	EMI shielding at 28.1 dB	(Weng et al. 2020)
Ti_3AlC_2	Ti_3C_2 MXene-xanthan	Vacuum-assisted filtration	NA	R.T.	2	Nanosheet	NA	50	Tensile strength at 121 MPa	(Sun, Ding, et al. 2021)
Ti_3AlC_2	PVDF-TrFE/ Ti_3C_2 MXene	Electro-spinning	DMF/ acetone	R.T.	1	Nanofiber	NA	2	P_{max} = 3.64 mW m^{-2}	(Wang, Shao, et al. 2021)

Yang et al. (Yang et al. 2020) fabricated the MXene incorporated with polyvinyl-alcohol/ sulfosuccinic acid (PVA/SSA) on polytetrafluoroethylene (PTFE) membrane through the solution film-coating or phase-inversion method using a membrane caster. Typically, 1.5 g PVA powder was dissolved in distilled water at 90 °C to form 1.5 wt% solution. Next, 20 wt% of SSA together with varying MXene contents were added dropwise to the PVA solution, forming the dope solution. A similar amount of distilled water was then added to dilute the dope solution to form the homogeneous mixture. The PVA/SSA/MXene solution underwent a degassing process under ultrasonication for 20 min prior to the membrane casting. The phase-inversion fabrication process of the MXene composite membrane is depicted in Figure 5.11. The as-fabricated MXene membrane exhibited high salt rejection at 99.8% as compared to that of bare PVA/PTFE membrane, which only achieved about 60% desalination. They explained that MXene had improved the interphase between the polymer, in addition to the enhancement in hydrophilicity of the polymer.

In another study, Alfahel et al. (Alfahel et al. 2020) prepared the crosslinked MXene/cellulose acetate (CA) composite membrane via a phase-inversion process using the doctor-blade knife. Different weights of delaminated $Ti_3C_2T_x$ MXene were dispersed in 10 mL of 1:1 acetone to acetic acid mixture and followed by 0.5 g of polyethylene glycol−400 (PEG−400). After mixing in the sonication bath for 60 min, a 1.5 g of CA was added to the suspension. The dope solution was then poured on the glass plate with doctor-blade knife to control the membrane thickness. The casted membrane was lastly cured in 15 °C water for 1 hour. For the chemical crosslinking process, the membrane was treated with acidic formaldehyde for 90 min at 50 °C. Pandey et al. (Pandey et al. 2020) also synthesized the crosslinked MXene/CA composite membrane via a phase-inversion technique. The main difference was that the crosslinked MXene/CA membrane prepared by Pandey et al. (Pandey et al. 2020) was treated at 60 °C for 60 min. Both studies showed that the

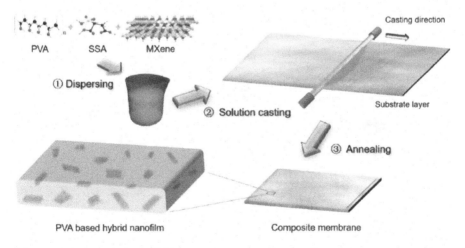

FIGURE 5.11   Fabrication process of $Ti_3C_2T_x$ MXene/PVA/SSA/PTFE membrane using a phase-inversion method (Yang et al. 2020).

MXene improved the hydrophilicity of the pristine CA membrane in addition to the higher water flux for improved desalination performance.

Apart from that, the supported-MXene composites can be prepared through the pressure-extrusion filtration approach. The carbonized cellulose microspheres@ void@$Ti_3C_2T_x$ MXene (CCM@void@MXene) composite film was fabricated via a pressure-extrusion filtration followed by mechanical pressing and annealing (Qian et al. 2021). The typical cellulose microsphere (CM) solution was added to MXene dispersion under constant stirring. The as-prepared mixture was filtrated under 2 MPa nitrogen gas pressure in a pressure-extrusion apparatus with polycarbonate membrane filter paper. The CM@MXene film was detached and vacuum-dried for 5 min at 60 °C. The dried CM@MXene composite film was put under 10 MPa mechanical pressing for 1 min accompanied by annealing in argon environment at 400 °C for 2 h to form CCM@void@MXene composite. Figure 5.12 illustrates the pressure-extrusion filtration route to fabricate the composite film.

Wang et al. (Wang, Shao, et al. 2021) developed the $Ti_3C_2T_x$ MXene/natural rubber composite with honeycomb structure via the filtration by a pressure-extrusion method. Typically, 1.5 g of natural rubber (NR) latex solution was dispersed with the addition of varying amount of $Ti_3C_2T_x$ MXene for 20 min. The homogenized solution was heated to 45 °C–50 °C and mixed with a dicumyl peroxide (DCP) curing solution, which was prepared by dissolving in 50 °C distilled water. The mixtures were then filtrated using a 0.4 μm filter membrane in the pressure-extrusion apparatus. Finally, the MXene/NR/DCP film was treated for vulcanization at 160 °C in a vacuum oven for 20 min. In their works, the cured and vulcanized MXene/NR composite exhibited the highest electromagnetic interference shielding (EMI) resistance at 63.5 dB and strengthening its tensile strength at 25.5 MPa.

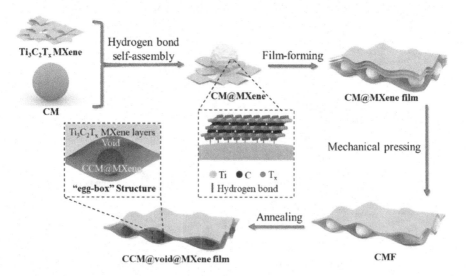

FIGURE 5.12 Schematic illustration of the preparation of CCM@void@MXene composite using a pressure-extrusion filtration (Qian et al. 2021).

The MXene/poly(diallymethylammonium chloride) (PDDA) composite through a simple chemical dropwise of PDDA solution to MXene suspension was fabricated and applied for the removal of perrhenate [Re(VII) or ReO_4^-] ions (Wang et al. 2019). In general, Ti_2CT_x MXene was prepared through the in situ HF produced from the LiF and HCl acid to etch the Al from Ti_2AlC MAX powder, stirring at 35 °C for 72 h. The delamination of MXene was done by adding the as-synthesized MXene to deionized water assisted with handshaking and centrifugation at 2000 rpm. For the preparation of Ti_2CT_x MXene nanosheet/PDDA composite (TCNS-P), 2 wt% of PDDA solution was added dropwise to 10 mg mL^{-1} TCNS colloidal solution and stirred under Ar environment for 2 h. The composite solid was collected via centrifugation at 10000 rpm and followed by freeze-drying for 2 days at −60 °C. They explained that the Re (VII) removal mechanism was due to the rapid adsorption of ReO_4^- ions on TCNS-P composite. The PDDA as a cationic polyelectrolyte exhibited strong electrostatic attraction of anionic ReO_4^-. The strong reductive potential of Ti_2CT_x MXene functional groups further reduced ReO_4^- to ReO_2.

Furthermore, Feng et al. (Feng et al. 2020) also constructed the MXene-derived layered double metal hydroxide (LDH) composite as an adsorbent in removal of nickel (Ni^{2+}) ions. In their investigations, the $Ti_3C_2T_x$ MXene was first functionalized by adding 2 g MXene into 5% sodium hydroxide (NaOH) for 24 h to form alkalized-MXene (alk-MXene). The alk-MXene/LDH composite was synthesized by mechanically stirring 0.5 g of alk-MXene and 0.5 g of LDH in 50 cm^3 of 5% NaOH for 24 hours. The LDH with positively charged surface underwent self-assembly deposition onto alk-MXene. The composite solid was collected via the centrifugation and finally dried at 60 °C. The adsorption capacity of pristine LDH (38.95 mg g^{-1}) was improved to 222.71 mg g^{-1} for that of alk-MXene/LDH. The enhancement reason given by them was due to the strong chemical bonding between functional group of adsorbent surfaces and Ni^{2+} ions, in addition to the high-level adsorption active sites. The hydroxyl (OH^-) functional groups of alk-MXene were chemically adsorbed Ni^{2+} ions and eventually precipitated as the nickel hydroxide.

Weng et al. (Weng et al. 2020) prepared aramid nanofiber (ANF)/MXene composite film via a simple vacuum-assisted filtration technique. In a typical synthesis process, ANF solution was prepared by dissolving 10 g Kevlar fiber into 1 L dimethyl sulfoxide (DMSO) with 10 g solid potassium hydroxide (KOH) and continuously stirring at room condition for 14 days. The as-prepared $Ti_3C_2T_x$ MXene was dispersed homogenously in deionized water and followed by mixing with ANF solution using the planetary vacuum mixer. The well-mixed solution was filtrated with a PTFE membrane filter with the vacuum-assisted system and dried in oven. Lastly, the ANF/MXene film was peeled off from PTFE membrane and cured in water to remove excess KOH for 24 h. In the same vein, Sun et al. (Sun, Ding, et al. 2021) studied the synthesis of MXene nanocomposite via a vacuum filtration process. In their reports, 0.03 wt% of xanthan solution was prepared under magnetic hotplate stirring for 6 h at 80 °C. After that, the MXene dispersion was added to the xanthan solution and stirred for another 2 h. The homogeneous solution was then filtrated through PVDF membrane filter and followed by drying at ambient condition. The obtained MXene-xanthan nanocomposite film was finally peeled from the PVDF filter.

A piezoelectric Ti_3C_2 MXene loaded with poly (vinylidene fluoride-trifluoride ethylene) (PVDF-TrFE) composite nanofibers was prepared using an electrospinning process (Wang, Shao, et al. 2021). The Ti_3C_2 MXene was first prepared based on their previous report (Zhao et al. 2019), accompanied with the delamination by sonication. The delaminated MXene was dispersed in N, N-dimethylformamide (DMF)/acetone solvent under sonication for an hour. Concurrently, PVDF-TrFE (PT) powder was dissolved in the DMF/acetone solvent for 1 h in another beaker. The mixture was added with different MXene dispersions, and the obtained composite suspension then underwent the electrospinning process on a static collector. The schematic illustration of the composite preparation through the electrospinning method is displayed in Figure 5.13a. They observed that a remarkable power density of 6.64 mW m^{-2} was obtained using the PT/MXene–2 wt% (PT/M2) nanofiber composite. The reason given for this result was that the narrow-size nanofibers MXene loaded to the PT enhanced the polarization performance. Their SEM images of the large-fiber PT and narrow-fiber PT/M2 are illustrated in Figure 5.13b, c.

In summary, various MXene-supported composites were investigated and fabricated via numerous techniques, including vacuum-assisted filtration, electrospinning, solution film casting with doctor-blade knife, simple magnetic stirring, and pressure-extrusion process. The addition of MXene as a support for other materials forming MXene composite had enhanced the properties of each substance. For instance, MXene-loaded on polymer membrane had improved the hydrophilicity of bare polymer membrane for the desalination. Due to the vast properties of MXene

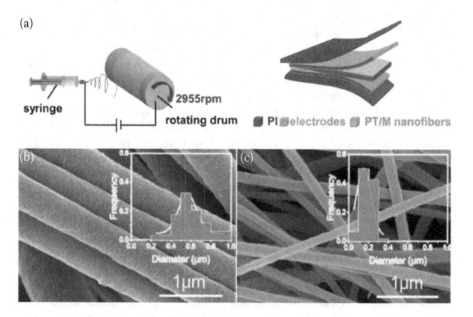

FIGURE 5.13   (a) Schematic illustration of the synthesis of PT/M composite using electrospinning technique. SEM images of (b) pristine PT and (c) PT/M nanofibers composite (Weng et al. 2020).

composite, it possesses high potential to be fabricated using various techniques in future investigation.

5.3 CONCLUSION AND PERSPECTIVES

In summary, this chapter recapitulated the recent progress associated with MXene-based photocatalysts for myriad preparation techniques, including ion-exchange, self-assembly, hydro- or solvothermal, sol-gel, calcination method, and so on. The typical MXene can serve as an effective electron-transport mediator for the electron-hole pairs separation in the composites. The Schottky junctions established by intimate coupling between MXene and a semiconductor can greatly promote the photogenerated electron-hole pairs separation, while the semiconductor continuously produced the charge carrier by harnessing the light exposure. Additionally, the 2D MXene structure can provide the large surface area, facilitate the pollutant to contact the active sites, boost the light-absorbing ability, and reduce the diffusion pathways for photogenerated charge carriers. These distinctive traits of the developed MXene-based photocatalysts have allowed them to serve as the most advantaged and encouraged novel candidates in the widespread applications, especially on the photocatalysis applications.

Despite the significant development in the MXene-based photocatalysts procedures thus far, there were some impediments and issues to be surmounted for improving the efficacy of a MXene-based photocatalyst system and deciphering the working mechanism behind the enhancement of the photocatalysis. Some future challenges and prospects in the progress of MXene-based photocatalysts are summarized as follows:

1. Exploration of more novel and effective synthetic methods is needed to strengthen the photo-heterojunction of the MXene-based photocatalysts for the ameliorated photocatalytic performance.
2. Investigation is needed on the intercorrelations among the MXene-based photocatalysts photoheterojunctions compositions, microstructures, and photocatalytic properties by using more systematic theoretical and computational measurements.
3. Theoretical modeling is essential to realize the physicochemical characteristics of MXene-based photocatalyst photo-heterojunctions and to explicate the relationship of the photocatalyst morphology, structural selectivity, and catalytic kinetics.
4. Despite the extensive applications of MXene-based photocatalysts reported so far, the correlation between the MXene structure and its photoactivity performance in the photoreaction were still not very clear.

ACKNOWLEDGMENTS

This research was supported by the Ministry of Higher Education of Malaysia (MoHE) through Fundamental Research Grant Scheme (FRGS/1/2019/TK02/UTAR/02/4). We also want to thank the Universiti Tunku Abdul Rahman (UTARRF/2020-C1/S04 and UTARRF/2020-C2/L02).

REFERENCES

Alfahel, Radwan, Reem S. Azzam, MhdAmmar Hafiz, et al. 2020. Fabrication of fouling resistant $Ti_3C_2T_x$ (MXene)/cellulose acetate nanocomposite membrane for forward osmosis application. *Journal of Water Process Engineering* 38:101551.

Baral, Basudev, K. Hemalata Reddy, and K. M. Parida. 2019. Construction of M-BiVO$_4$/T-BiVO$_4$ isotype heterojunction for enhanced photocatalytic degradation of Norfloxacine and Oxygen evolution reaction. *Journal of Colloid and Interface Science* 554:278–295.

Chen, Wei, Rui-Qiang Yan, Gui-Hua Chen, Man-Yu Chen, Guo-Bo Huang, and Xiao-Heng Liu. 2019. Hydrothermal route to synthesize helical $CdS@ZnIn_2S_4$ core-shell heterostructures with enhanced photocatalytic hydrogeneration activity. *Ceramics International* 45 (2, Part A):1803–1811.

Cheng, Lei, Xin Li, Huaiwu Zhang, and Quanjun Xiang. 2019. Two-dimensional transition metal MXene-based photocatalysts for solar fuel generation. *The Journal of Physical Chemistry Letters* 10 (12):3488–3494.

Cui, Ce, Ronghui Guo, Erhui Ren, et al. 2021. Facile hydrothermal synthesis of rod-like Nb_2O_5/Nb_2CT_x composites for visible-light driven photocatalytic degradation of organic pollutants. *Environmental Research* 193:110587.

Cui, Ce, Ronghui Guo, Hongyan Xiao, et al. 2020. Bi_2WO_6/Nb_2CT_x MXene hybrid nanosheets with enhanced visible-light-driven photocatalytic activity for organic pollutants degradation. *Applied Surface Science* 505:144595.

Du, Xin, Tianyu Zhao, Ziyuan Xiu, et al. 2020. $BiVO_4@ZnIn_2S_4/Ti_3C_2$ MXene quantum dots assembly all-solid-state direct Z-Scheme photocatalysts for efficient visible-light-driven overall water splitting. *Applied Materials Today* 20:100719.

Feng, Xiaofang, Zongxue Yu, Runxuan Long, et al. 2020. Self-assembling 2D/2D (MXene/LDH) materials achieve ultra-high adsorption of heavy metals Ni^{2+} through terminal group modification. *Separation and Purification Technology* 253:117525.

Feng, Xiaofang, Zongxue Yu, Yuxi Sun, et al. 2021. Review MXenes as a new type of nanomaterial for environmental applications in the photocatalytic degradation of water pollutants. *Ceramics International* 47 (6):7321–7343.

Gao, Wanguo, Xiaoman Li, Shijian Luo, et al. 2021. In situ modification of cobalt on MXene/TiO$_2$ as composite photocatalyst for efficient nitrogen fixation. *Journal of Colloid and Interface Science* 585:20–29.

Gopinath, Kannappan Panchamoorthy, Nagarajan Vikas Madhav, Abhishek Krishnan, Rajagopal Malolan, and Goutham Rangarajan. 2020. Present applications of titanium dioxide for the photocatalytic removal of pollutants from water: A review. *Journal of Environmental Management* 270:110906.

Guo, Zhonglu, Jian Zhou, Linggang Zhu, and Zhimei Sun. 2016. MXene: A promising photocatalyst for water splitting. *Journal of Materials Chemistry A* 4 (29):11446–11452.

Hao, Chongyan, Yuan Liao, Yang Wu, et al. 2020. RuO_2-loaded TiO_2–MXene as a high performance photocatalyst for nitrogen fixation. *Journal of Physics and Chemistry of Solids* 136:109141.

Hieu, Vu Quang, Thanh Khoa Phung, Thanh-Quang Nguyen, et al. 2021. Photocatalytic degradation of methyl orange dye by Ti_3C_2–TiO_2 heterojunction under solar light. *Chemosphere* 276:130154.

Hong, Long-fei, Rui-tang Guo, Ye Yuan, et al. 2020. Recent progress of two-dimensional MXenes in photocatalytic applications: A review. *Materials Today Energy* 18:100521.

Huang, Huoshuai, Xin Jiang, Najun Li, et al. 2021. Noble-metal-free ultrathin MXene coupled with In_2S_3 nanoflakes for ultrafast photocatalytic reduction of hexavalent chromium. *Applied Catalysis B: Environmental* 284:119754.

Huang, Kelei, Chunhu Li, and Xiangchao Meng. 2020. In-situ construction of ternary Ti_3C_2 MXene@TiO_2/$ZnIn_2S_4$ composites for highly efficient photocatalytic hydrogen evolution. *Journal of Colloid and Interface Science* 580:669–680.

Im, Jong Kwon, Erica Jungmin Sohn, Sewoon Kim, et al. 2021. Review of MXene-based nanocomposites for photocatalysis. *Chemosphere* 270:129478.

Jaya Prakash, Niranjana, and Balasubramanian Kandasubramanian. 2021. Nanocomposites of MXene for industrial applications. *Journal of Alloys and Compounds* 862:158547.

Khan, Azmat Ali, and Muhammad Tahir. 2021. Well-designed 2D/2D Ti_3C_2TA/R MXene coupled g-C_3N_4 heterojunction with in-situ growth of anatase/rutile TiO_2 nucleates to boost photocatalytic dry-reforming of methane (DRM) for syngas production under visible light. *Applied Catalysis B: Environmental* 285:119777.

Kuang, Panyong, Jingxiang Low, Bei Cheng, Jiaguo Yu, and Jiajie Fan. 2020. MXene-based photocatalysts. *Journal of Materials Science & Technology* 56:18–44.

Li, Jiajia, Kai Han, Jinhua Huang, et al. 2021. Polarized nucleation and efficient decomposition of Li_2O_2 for Ti_2C MXene cathode catalyst under a mixed surface condition in lithium-oxygen batteries. *Energy Storage Materials* 35:669–678.

Li, Tengfei, Xudong Yan, Lujun Huang, et al. 2019. Fluorine-free $Ti_3C_2T_x$ (T = O, OH) nanosheets (~50–100 nm) for nitrogen fixation under ambient conditions. *Journal of Materials Chemistry A* 7 (24):14462–14465.

Li, Xing, Yang Bai, Xian Shi, et al. 2021. Mesoporous g-C_3N_4/MXene (Ti_3C_2Tx) heterojunction as a 2D electronic charge transfer for efficient photocatalytic CO_2 reduction. *Applied Surface Science* 546:149111.

Liu, Anmin, Qiyue Yang, Xuefeng Ren, et al. 2020. Energy- and cost-efficient NaCl-assisted synthesis of MAX-phase Ti_3AlC_2 at lower temperature. *Ceramics International* 46 (5):6934–6939.

Liu, Huanhuan, Cai Yang, Xingyun Jin, Junbo Zhong, and Jianzhang Li. 2020. One-pot hydrothermal synthesis of MXene Ti_3C_2/TiO_2/BiOCl ternary heterojunctions with improved separation of photoactivated carries and photocatalytic behavior toward elimination of contaminants. *Colloids and Surfaces: A Physicochemical and Engineering Aspects* 603:125239.

Liu, Ruiting, Miao Miao, Yahui Li, Jianfeng Zhang, Shaomei Cao, and Xin Feng. 2018. Ultrathin biomimetic polymeric $Ti_3C_2T_x$ MXene composite films for electromagnetic interference shielding. *ACS Applied Materials & Interfaces* 10 (51):44787–44795.

Liu, Xiaoyan, and Chuansheng Chen. 2020. Mxene enhanced the photocatalytic activity of ZnO nanorods under visible light. *Materials Letters* 261:127127.

Liu, Yunpeng, Yu-Hang Li, Xiaoyao Li, et al. 2020. Regulating electron–hole separation to promote photocatalytic H2 evolution activity of nanoconfined Ru/MXene/TiO_2 catalysts. *ACS Nano* 14 (10):14181–14189.

Low, Jingxiang, Liuyang Zhang, Tong Tong, Baojia Shen, and Jiaguo Yu. 2018. TiO_2/MXene Ti_3C_2 composite with excellent photocatalytic CO_2 reduction activity. *Journal of Catalysis* 361:255–266.

Lu, Siyi, Ge Meng, Can Wang, and Hong Chen. 2021. Photocatalytic inactivation of airborne bacteria in a polyurethane foam reactor loaded with a hybrid of MXene and anatase TiO_2 exposing {001} facets. *Chemical Engineering Journal* 404:126526.

Mei, Jun, Godwin A. Ayoko, Chunfeng Hu, John M. Bell, and Ziqi Sun. 2020. Two-dimensional fluorine-free mesoporous Mo_2C MXene via UV-induced selective etching of Mo_2Ga_2C for energy storage. *Sustainable Materials and Technologies* 25:e00156.

Natu, Varun, Rahul Pai, Maxim Sokol, Michael Carey, Vibha Kalra, and Michel W. Barsoum. 2020. 2D Ti3C2Tz MXene synthesized by water-free etching of Ti3AlC2 in polar organic solvents. *Chem* 6 (3):616–630.

Pandey, Ravi P., P. Abdul Rasheed, Tricia Gomez, Reem S. Azam, and Khaled A. Mahmoud. 2020. A fouling-resistant mixed-matrix nanofiltration membrane based on covalently cross-linked $Ti_3C_2T_X$ (MXene)/cellulose acetate. *Journal of Membrane Science* 607:118139.

Qian, Kunpeng, Qianfan Zhou, Hongmin Wu, et al. 2021. Carbonized cellulose microsphere@ void@MXene composite films with egg-box structure for electromagnetic interference shielding. *Composites Part A: Applied Science and Manufacturing* 141:106229.

Shi, Qunrong, Xiaoyue Zhang, Yong Yang, et al. 2021. 3D hierarchical architecture collaborating with 2D/2D interface interaction in $NiAl-LDH/Ti_3C_2$ nanocomposite for efficient and selective photoconversion of CO_2. *Journal of Energy Chemistry* 59:9–18.

Song, Limin, and Shujuan Zhang. 2010. A simple mechanical mixing method for preparation of visible-light-sensitive NiO–CaO composite photocatalysts with high photocatalytic activity. *Journal of Hazardous Materials* 174 (1):563–566.

Sun, Bin, Furong Tao, Zixuan Huang, et al. 2021. Ti_3C_2 MXene-bridged Ag/Ag_3PO_4 hybrids toward enhanced visible-light-driven photocatalytic activity. *Applied Surface Science* 535:147354.

Sun, Yan, Ruonan Ding, Sung Yong Hong, et al. 2021. MXene-xanthan nanocomposite films with layered microstructure for electromagnetic interference shielding and Joule heating. *Chemical Engineering Journal* 410:128348.

Ta, Qui Thanh Hoai, Nghe My Tran, and Jin-Seo Noh. 2020. Rice crust-like $ZnO/Ti_3C_2T_x$ MXene hybrid structures for improved photocatalytic activity. *Catalysts* 10 (10):1140.

Wang, He, Lei Chen, Yanping Sun, et al. 2021. Ti_3C_2 Mxene modified $SnNb_2O_6$ nanosheets Schottky photocatalysts with directed internal electric field for tetracycline hydrochloride removal and hydrogen evolution. *Separation and Purification Technology* 265:118516.

Wang, Lin, Huan Song, Liyong Yuan, et al. 2019. Effective removal of anionic Re(VII) by surface-modified Ti_2CT_x MXene nanocomposites: Implications for Tc(VII) sequestration. *Environmental Science & Technology* 53 (7):3739–3747.

Wang, Shan, He-Qing Shao, Yong Liu, et al. 2021. Boosting piezoelectric response of PVDF-TrFE via MXene for self-powered linear pressure sensor. *Composites Science and Technology* 202:108600.

Weng, Chuanxin, Tianle Xing, Hao Jin, et al. 2020. Mechanically robust ANF/MXene composite films with tunable electromagnetic interference shielding performance. *Composites Part A: Applied Science and Manufacturing* 135:105927.

Wu, Jinlei, Yu Zhang, Po Lu, et al. 2021. Engineering 2D multi-hetero-interface in the well-designed nanosheet composite photocatalyst with broad electron-transfer channels for highly-efficient solar-to-fuels conversion. *Applied Catalysis B: Environmental* 286:119944.

Yang, Guang, Zongli Xie, Aaron W. Thornton, et al. 2020. Ultrathin poly (vinyl alcohol)/ MXene nanofilm composite membrane with facile intrusion-free construction for pervaporative separations. *Journal of Membrane Science* 614:118490.

Zeng, Gongchang, Ying Cao, Yixiao Wu, et al. 2021. $Cd0.5Zn0.5S/Ti_3C_2$ MXene as a Schottky catalyst for highly efficient photocatalytic hydrogen evolution in seawater. *Applied Materials Today* 22:100926.

Zhang, Huoli, Man Li, Changxin Zhu, Qingjie Tang, Peng Kang, and Jianliang Cao. 2020. Preparation of magnetic $\alpha-Fe_2O_3/ZnFe_2O_4@Ti_3C_2$ MXene with excellent photocatalytic performance. *Ceramics International* 46 (1):81–88.

Zhang, Junzheng, Jingjing Shi, Sheng Tao, Lei Wu, and Jun Lu. 2021. Cu_2O/Ti_3C_2MXene heterojunction photocatalysts for improved CO_2 photocatalytic reduction performance. *Applied Surface Science* 542:148685.

Zhao, Sen, Ravi Nivetha, Yu Qiu, and Xiaohui Guo. 2020. Two-dimensional hybrid nanomaterials derived from MXenes ($Ti_3C_2T_x$) as advanced energy storage and conversion applications. *Chinese Chemical Letters* 31 (4):947–952.

Zhao, Xing, Xiang-Jun Zha, Jun-Hong Pu, et al. 2019. Macroporous three-dimensional MXene architectures for highly efficient solar steam generation. *Journal of Materials Chemistry A* 7 (17):10446–10455.

Zhao, Yang, Meidan Que, Jin Chen, and Chunli Yang. 2020. MXenes as co-catalysts for the solar-driven photocatalytic reduction of CO_2. *Journal of Materials Chemistry C* 8 (46):16258–16281.

Zou, Xue, Xuesong Zhao, Jiaxing Zhang, Wei Lv, Ling Qiu, and Zhenghua Zhang. 2021. Photocatalytic degradation of ranitidine and reduction of nitrosamine dimethylamine formation potential over MXene–Ti_3C_2/MoS_2 under visible light irradiation. *Journal of Hazardous Materials* 413:125424.

Zu, Daoyuan, Haoran Song, Yuwei Wang, et al. 2020. One-pot in-situ hydrothermal synthesis of CdS/Nb_2O_5/Nb_2C heterojunction for enhanced visible-light-driven photodegradation. *Applied Catalysis B: Environmental* 277:119140.

6 Application of MXene-Based Photocatalyst for Photocatalytic Water-Splitting

Liuyun Chen, Tongming Su, and Zuzeng Qin
School of Chemistry and Chemical Engineering, Guangxi University

Hongbing Ji
Fine Chemical Industry Research Institute, School of Chemistry, Sun Yat-Sen University

CONTENTS

6.1 INTRODUCTION

In the past decades, both the energy crisis and environmental pollution were serious problems facing humans. Notably, fossil fuel, the most needed energy source in the world, was neither renewable nor environmentally friendly. Therefore, the development of a new renewable clean energy source is needed to alleviate the pressure on energy and the environment (Xin et al. 2020). Solar energy was inexhaustible and had been considered by the researchers. Fox example, photocatalysis is a process of converting solar energy into chemical energy (Boyjoo et al. 2017). Among them, the photocatalytic water-splitting for hydrogen-evolution reaction is promising to solve the energy and environmental crisis (Cao et al. 2020; Chen, Huang, et al. 2020). Photocatalytic water-splitting is a process of reducing H^+ to H_2 by the electrons generated in the photocatalyst under light irradiation (Kumaravel et al. 2019; Chen, Xie, et al. 2021). To realize the photocatalytic water-splitting reaction, it is necessary to meet the following conditions. Firstly, under light irradiation, the energy absorbed by the photocatalyst must larger than or equal to the band gap of the photocatalyst. Secondly, photogenerated electrons can rapidly transfer to the active site on the surface of the photocatalyst. Thirdly, the conduction band potential of the photocatalyst should be more negative than that of the hydrogen standard electrode (0 V vs Normal hydrogen electrode (NHE) at pH = 0). Therefore, an ideal photocatalyst should have these corresponding advantages of visible light absorption capacity, excellent separation capability of photo-generated electrons and holes, and enough negative conduction potential.

MXene was two-dimensional (2D) transition metal carbide, nitride, and carbonitride, which was discovered in 2011 (Naguib et al. 2011). $M_{n+1}X_nT_x$ was the general formula for MXene materials. M stands for transition metal (e.g. Sc, Ti, Zr, Hf, V, and Nb.), X was for carbon or nitrogen and T stands for the termination group (F-termination, O-termination, and OH-termination) on the surface (Alhabeb et al. 2017). MXene generally presents an accordion-like multilayer 2D structure or monolayer nanosheets (Su et al. 2019a). With a hydrophilic surface and high electronic conductivity, MXene had the potential for energy storage (Liu, Yu, et al. 2020), conduction (Tang et al. 2019), medical treatment (Arabi Shamsabadi et al. 2018), and catalysis (Kuang et al. 2020). In the field of photocatalysis, MXene was mainly applied in photocatalytic hydrogen evolution as a co-catalyst (Xiao et al. 2020). The excellent electrical conductivity of MXene enables it to act as a co-catalyst to capture photoinduced electrons on the photocatalyst in the photocatalytic process (Tang et al. 2020). For example, Ti_3C_2 and Nb_2C MXene were applied in photocatalytic hydrogen-evolution reaction as co-catalysts (Su, Peng, et al. 2018). In addition, the termination of MXene can act as the active sites to reduce H_2O to H_2 in the photocatalytic water-splitting process. MXene can be prepared by etching the MAX phase. MXene with multilayer and monolayer 2D structure structure can be obtained with different etchants and dispersants. For the photocatalytic hydrogen-evolution reaction, the 2D structure of photocatalyst shows the following advantages (Su, Shao, et al. 2018): (I) the band gap of photocatalyst can be tuned by adjusting the number of layers and the thickness of nanosheets. (II) The distance from the inside to the surface was shortened. (III) The interface between the semiconductors was larger and closer. (IV) More active sites can be exposed on the surface of the photocatalyst.

MXene with different terminal groups (F-, O-, OH-) can be prepared using different etching methods (Ran et al. 2017). MXene-based photocatalysts are usually composed of MXene and semiconductors with MXene as the co-catalyst. The MXene-based photocatalysts for photocatalytic water-splitting can be divided into three categories. (I) The metal oxides derived from transition metal atoms can be formed on the surface of MXene via the oxidation treatment (Wu et al. 2020; Xu, Yang, et al. 2020). (II) A semiconductor is generated in situ on a MXene surface with the external additive precursor (Ai et al. 2020; Fang et al. 2020). (III) MXene was combined with the prefabricated semiconductor (Li, Lu, et al. 2020; Liu, Lu, et al. 2020). After the coupling of the MXene and the photocatalyst, a Schottky junction can be formed due to the suitable Fermi level of MXene, enabling the efficient separation of photogenerated electrons and holes. Therefore, MXene as a co-catalyst has a great prospect in the field of photocatalytic water-splitting.

6.2 MXENE-BASED PHOTOCATALYSTS FOR PHOTOCATALYTIC WATER-SPLITTING

MXene cannot produce photoinduced electrons and holes under light irradiation due to its metallic properties (Huang et al. 2020; Ji et al. 2021); it was generally used as a co-catalyst in the photocatalytic hydrogen-evolution reaction. Different MXene-based photocatalysts can be obtained by different preparation methods, and MXene also plays a different role with different photocatalysts. At present, the preparation methods of MXene-based photocatalysts can be summarized as partial oxidation, complete oxidation, and coupling with semiconductors. In this case, the photocatalytic activity of the semiconductors can be enhanced by using MXene as the co-catalyst.

6.2.1 MXENE-BASED PHOTOCATALYSTS PREPARED BY INCOMPLETE OXIDATION OF MXENES

The incomplete oxidation method was used to oxidize the metal atoms on the surface of MXene into metal oxides to prepare the metal oxide/MXene composite catalyst (Wang, Lu, et al. 2020). The most common type of photocatalyst composite is the TiO_2/Ti_3C_2; it used Ti atoms on the surface of Ti_3C_2 as the titanium source, the Ti atoms were partially oxidized and converted into TiO_2 as the photocatalyst by hydrothermal or calcination method, and the remaining Ti_3C_2 was used as the co-catalyst (Low et al. 2018). In this way, MXene was used as the template to construct the photocatalyst composite with the layered structure.

Ti_3C_2 was one of the most widely used MXene in the field of photocatalysis due to the fact that Ti atoms on its surface can be oxidized to TiO_2 (Cheng et al. 2019). TiO_2 is the earliest semiconductor used for photocatalytic water-splitting in the world (Fujishima and Honda 1972), which is also the most widely studied semiconductor in the field of photocatalysis (Xing et al. 2018; Chen, Cheng, et al. 2020; Qin, Chen, et al. 2020). The typical synthesis scheme of $1T\text{-}WS_2@TiO_2@Ti_3C_2$ composite was shown in Figure 6.1a (Li, Ding, et al. 2019). $TiO_2@Ti_3C_2$ was

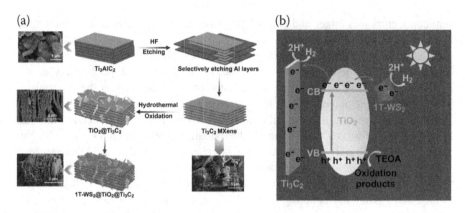

FIGURE 6.1 (a) Schematic illustration of the preparation of $1T\text{-}WS_2@TiO_2@Ti_3C_2$ composites (SEM images of Ti_3C_2, $TiO_2@Ti_3C_2$, and $1T\text{-}WS_2@TiO_2@Ti_3C_2$ were shown in the inset). (b) Schematic of the photocatalytic mechanism of the $1T\text{-}WS_2@TiO_2@Ti_3C_2$ composites (Li, Ding, et al. 2019).

obtained by HF etching and the hydrothermal-oxidation method. The SEM image of $TiO_2@Ti_3C_2$ shows that TiO_2 nanosheets were randomly distributed on the surface of Ti_3C_2. In addition, the TiO_2 nanosheets grew through the Ti_3C_2 layer, which caused by the addition of $NaBF_4$ and HCl during hydrothermal oxidation. $NaBF_4$ and HCl were used as the morphology control agents of TiO_2 nanosheets. After the formation of the $TiO_2@Ti_3C_2$ composites, $1T\text{-}WS_2$ was generated in situ on $TiO_2@Ti_3C_2$ with WCl_6 and thioacetamide as the precursor (Yi et al. 2018). As shown in Figure 6.1b, the photogenerated electrons on the TiO_2 conduction band (CB) were transferred to Ti_3C_2 and $1T\text{-}WS_2$ for the reduction reaction. The TiO_2 nanosheet generated in situ on the MXene exhibited more contact interfaces with and Ti_3C_2 and $1T\text{-}WS_2$, which is beneficial for the charge carriers transfer. Moreover, the flat and flaky structure of TiO_2 was conducive to the rapid transfer of photoinduced electrons on the surface. In addition, $1T\text{-}WS_2$ nanoparticles increased the specific surface area of the photocatalyst and provide more active sites for photocatalytic hydrogen evolution (Manikandan et al. 2018). In general, the extraordinary conductivity of Ti_3C_2 and $1T\text{-}WS_2$ greatly enhanced the electron-transfer capability, thus achieving efficient photogenerated charge separation and enhanced photocatalytic performance (Li, Ding, et al. 2019).

A similar study was carried out by the other researchers. With the $NaBF_4$ and HCl as morphology control agents, TiO_2/Ti_3C_2 type photocatalysts was obtained (Peng et al. 2016; Peng et al. 2018). Under the action of HCl and $NaBF_4$, TiO_2 nanosheets were grown in situ by hydrothermal oxidation on the Ti_3C_2 surface. In another hydrothermal reaction, $Na_2MoO_4\cdot2H_2O$ was used as the Mo source to prepare $Ti_3C_2@TiO_2@MoS_2$. TiO_2 nanosheets grew through the Ti_3C_2 layer, and MoS_2 was evenly distributed on the $Ti_3C_2@TiO_2$ surface. With the assistance of $NaBF_4$, due to the low energy of the (101) facets of adsorption F-, the growth of the (001) facets with high energy can be promoted during the growth of TiO_2 crystals (Zhou et al. 2013; Li et al. 2018). By using this method, TiO_2 nanosheets interspersed in the Ti_3C_2 layer and exposed several (001) planes. The most active (001) facets of TiO_2 nanosheet cross the 2D Ti_3C_2. In addition, the tight contact between the two phases can promote the separation of photogenerated carriers on the (001) plane of TiO_2, thus improving the photocatalytic activity (Chen et al. 2016). The photocatalytic hydrogen-evolution reaction mechanism for $Ti_3C_2@TiO_2@MoS_2$ was proposed. The exposed (101) and (001) facets of TiO_2 nanosheets can form the surface heterojunction within a single TiO_2 particle. Through the weak van der Waals interaction, a lot of MoS_2 nanosheets were immobilized on the (101) surface of TiO_2. The independent sandwich Mo-S-Mo resulted in a large exposure of the Mo-edge with its metallic properties and high d-electron density (Guo et al. 2016), which provide a large number of photocatalytic active sites. This system has three transfer paths of photoinduced electrons. (I) Electrons transferred from the CB of TiO_2 (001) planes to Ti_3C_2. (II) Electrons transferred from the CB of TiO_2 (001) plane to TiO_2 (101) plane and then to MoS_2. (III) Electrons transferred from the CB of TiO_2 (101) to MoS_2. The faster electron transfer greatly improved the photocatalytic hydrogen production activity (Liu et al. 2019).

In addition to Ti_3C_2, the Nb_2C MXene can also be applied to photocatalytic hydrogen production as a co-catalyst (Su, Peng, et al. 2018). Nb_2O_5 was a promising semiconductor photocatalyst due to its excellent stability (Qu et al. 2020; Xu, Jiang, et al. 2020). However, as a metal oxide, Nb_2O_5 suffered from the recombination of photogenerated electrons and holes (Kulkarni et al. 2018). When Nb_2C was calcined in CO_2, part of the surface Nb atoms were partially oxidized, and Nb_2O_5 was grown in situ on the surface Nb_2C (Figure 6.2). In this Nb_2O_5/Nb_2C system, a tight heterojunction was formed between Nb_2C and Nb_2O_5, which effectively promoted the separation of photoproduction electrons and holes (Yan et al. 2014). Interestingly, the photocatalytic activity of physically mixed Nb_2O_5/Nb_2C was lower than the Nb_2O_5/Nb_2C composites obtained by the one-step CO_2 oxidation. In this mild CO_2 oxidation process, Nb atoms are oxidized to Nb_2O_5, and the remaining carbon were not oxidized but transformed into disordered carbon; thus, the $Nb_2O_5/C/Nb_2C$ composite was formed. For the $Nb_2O_5/C/Nb_2C$, the disordered carbon can be regarded as the binder of Nb_2C and Nb_2O_5, which maintained the structural stability of Nb_2C and Nb_2O_5 and created the electron-transfer channel, which was beneficial to the separation of photogenerated electrons and holes. Moreover, the Schottky barrier formed between the Nb_2C and Nb_2O_5 was considered as an electron trap, which further enhanced the separation of the photogenerated electron-hole pairs (Tian et al. 2019).

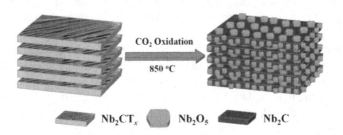

FIGURE 6.2 Schematic of the one-step CO_2 oxidation process (Su, Peng, et al. 2018).

6.2.2 Carbon Doped TiO₂ Photocatalyst Prepared by Oxidation of MXene

TiO_2 can only absorb ultraviolet light due to its wide band gap, though it is the most widely used metal-oxide semiconductor (Sun, Yuan, et al. 2020). The enhancement of the light-absorption capacity of TiO_2 was the key to realize the practical application of TiO_2 in the field of photocatalysis (Sun et al. 2017). Doping with other metallic or nonmetallic atoms is an effective way to improve the photocatalytic performance of the TiO_2. For example, the controlled and intentional carbon doping is a common and effective strategy to increase photocatalytic hydrogen evolution activity of TiO_2 under visible light (Kong et al. 2020). The method of complete oxidation can oxidize MXene to form the carbon doped metal oxide and maintain the original multilayer structure or nanosheet structure.

High carbon doping would introduce a new impurity energy level at the valence band of the photocatalyst, thereby promoting the effective separation of photo-generated charge carriers and reducing the bandgap of the photocatalyst, thus greatly improving the utilization of light (Shao et al. 2017). The addition of the new impurity energy level was attributed to the carboxylate groups from the in situ carbon doping with a strong electron absorption (Ganesh et al. 2018). Different from other annealing processes at high temperature, HC-TiO₂ (highly carbon doped TiO_2) was prepared by hydrothermal oxidation at low temperature with MXene as the precursor (Jia et al. 2018). The HC-TiO₂ nanorods produced from MXene were tightly assembled to form a multangular structure. The UV-vis spectra of HC-TiO₂ and P25 showed that the HC-TiO₂ exhibited a narrower band gap than P25 (commercial TiO_2). The flat band potential of HC-TiO₂ was more negative than that of P25, and the more negative flat band potential indicates the stronger reducing capacity in the conduction band of the photocatalyst, which is beneficial for the photocatalytic reaction (Xiong et al. 2019). The photocatalytic mechanism of HC-TiO₂ was shown in Figure 6.3. The one-dimensional structure of TiO_2 nanorods makes the photogenerated charge carriers mainly move in an axial direction. The diffusion length of photogenerated charge carriers was reduced by the formation of HC-TiO₂ nanorods, with layered valence bands after highly carbon doping. Due to the impurities formed by carbon doping, the acceptor potential of TiO_2 is higher than the maximum value of valence band and can be used as an effective hole-transfer medium (Zhang et al. 2018), which effectively facilitated the separation of photogenerated electrons and holes. The photocatalytic hydrogen-evolution rates of carbon doped TiO_2 with different carbon sources were shown in Table 6.1, indicating that Ti_3C_2 can be used as a carbon source to prepare high-efficiency visible light C-TiO₂ photocatalyst.

Nonmetallic doping can reduce the band gap energy of TiO_2 and enhance the light-absorption capacity of TiO_2 and the separation of photoinduced electron-hole pairs in TiO_2 (Qin, Li, et al. 2020; Wang, Rao, et al. 2020). In addition to carbon doping (Xu, Sun, et al. 2018), sulfur doping (Irandost et al. 2019) can also form the sulfur dopant level with potential higher than the valance band of TiO_2, thus enhancing the photocatalytic activity. Ti_3C_2 can be slowly oxidized to TiO_2 nanoparticles and layered carbon substrate in the CO_2 atmosphere at high temperature

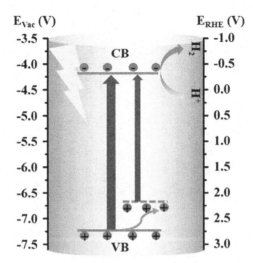

FIGURE 6.3 Schematic of photocatalytic mechanism for HC-TiO$_2$. CB = conduction band, VB = valence band (Jia et al. 2018).

(Yuan et al. 2018). Layered carbon substrate usually has a large specific surface area and excellent conductivity and can be regarded as a co-catalyst (Sun, He, et al. 2020). As shown in Figure 6.4a, Ti$_3$C$_2$ after sulfur impregnation was slowly oxidized in the CO$_2$ atmosphere to produce sulfur-doped TiO$_2$ nanoparticles, which was fixed on the laminated carbon substrate and formed the laminated defect-controlled sulfur doped TiO$_2$/C (LDC-S-TiO$_2$/C). By further oxidation in the air, the carbon content covered on the surface of TiO$_2$ was reduced, while more defects were formed in the laminated carbon substrate. The defect was considered as the active site of photocatalytic hydrogen evolution (Yin et al. 2019). LDC-S-TiO$_2$/C showed two band gaps; one was consistent with the LDC-TiO$_2$ (2.88 eV) and the other is attributed to the sulfur dopant level 1.62 eV. The influence of S doping on the band structure of TiO$_2$ was further studied by DFT calculation. Supercells of rutile TiO$_2$ and sulfur-atom doped TiO$_2$ were respectively constructed for research. The valance band of the sulfur doped TiO$_2$ exhibited significant positive shift, and the band gap of sulfur atom doped TiO$_2$ was shifted to 1.82 eV. The mechanism of photocatalytic hydrogen evolution of LDC-S-TiO$_2$/C is shown in Figure 6.4b. When S-TiO$_2$ absorbs visible light, electrons can be excited from the valance band to the conduction band of TiO$_2$ or from the sulfur-doping level to CB. Due to the high conductivity of laminated carbon and its close contact with TiO$_2$, photogenerated electrons can be transferred to laminated carbon rapidly. Moreover, electrons rapidly transferred to the defect, which was the active site, to participate in the reduction reaction.

Graphitic carbon nitride (g-C$_3$N$_4$) was a typical 2D metal-free organic polymer semiconductor with a suitable band edge (2.8 eV) for the absorption of visible light (Wang et al. 2009). The simple preparation method and ultrathin nanosheet structure of g-C$_3$N$_4$ have attracted great attention (Cai et al. 2020; Si et al. 2020). Carbon doped TiO$_2$/g-C$_3$N$_4$ photocatalysts were prepared by a one-step and

TABLE 6.1

Photocatalytic Hydrogen Evolution of Carbon Doped TiO$_2$ with Different Carbon Sources

Photocatalyst	Source of C-Doped	Amount of Catalyst (mg)	Sacrificial Reagent	Light Condition	H$_2$ Production Rate (μmol·g^{-1}·h^{-1})	Multiple of Enhance *	References
TiO$_2$/C	Ti$_3$C$_2$	50	10 vol% TEOA	300 W Xe lamp (420 nm cutoff filter)	69.00	3.0 times (TiO$_2$)	(Wang, Shen, et al. 2020)
HC-TiO$_2$	Ti$_3$C$_2$	20	10 vol% TEOA	300 W Xe lamp	33.04	9.7 times (TiO$_2$)	(Jia et al. 2018)
C-TiO2/g-C$_3$N$_4$	Ti$_3$C$_2$	10	10 vol% TEOA	300 W Xe lamp (420 nm cutoff filter)	1409.00	8.0 times (g-C$_3$N$_4$) 24.0 times (TiO$_2$)	(Han et al. 2020)
C-TiO$_2$/g-C$_3$N$_4$	TiC	50	10 vol% TEOA	300 W Xe lamp (420 nm cutoff filter)	1146.00	/	(Yang, Qin, et al. 2017)
TiC@C-TiO$_2$	TiC	40	16 vol% methanol	300 W Xe lamp (420 nm cutoff filter)	279.23	30.2 times (TiO$_2$)	(Yang, Zhang, et al. 2017)
C-TiO$_2$@g-C$_3$N$_4$	carbon sphere	10	10 vol% methanol	300 W Xe lamp (420 nm cutoff filter)	356	10.5 times (g-C$_3$N$_4$) 22.7 times (C-TiO$_2$)	(Zou, Shi, Ma, Fan, Lu, et al. 2017)
g-C$_3$N$_4$/Au/C-TiO$_2$	carbon spheres	10	10 vol% TEOA	300 W Xe lamp (420 nm cutoff filter)	129	42.0 times (g-C$_3$N$_4$) 86.0 times (C-TiO$_2$)	(Zou, Shi, Ma, Fan, Niu, et al. 2017)

* The ratio of the photocatalytic hydrogen production rate of the composite to that of the pure photocatalyst (in brackets).

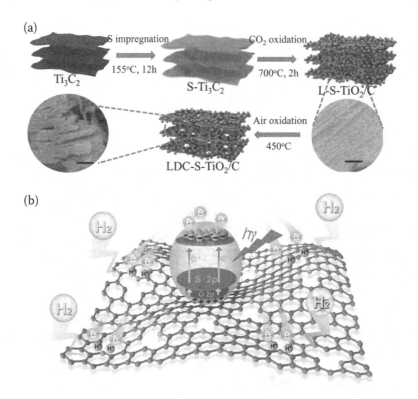

FIGURE 6.4   (a) Schematic diagram of the synthesis of LDC-S-TiO$_2$/C. (The SEM images of L-S-TiO$_2$/C and LDC-S-TiO$_2$/C was shown in inset). L = laminated, LCD = laminated defect controlled. (b) Mechanism of photocatalytic Hydrogen evolution of LDC-S-TiO$_2$/C, in which the carbon atoms are marked in blue and the sulfur atoms are marked in red (Yuan et al. 2018).

simple heat-treatment method (Han et al. 2020). As shown in Figure 6.5a, C-TiO$_2$/g-C$_3$N$_4$ was prepared by simple calcination of the multilayer Ti$_3$C$_2$ and the bulk g-C$_3$N$_4$ mixture. Interestingly, after the heat treatment, 2D/2D heterojunction between carbon-doped TiO$_2$ nanosheets and g-C$_3$N$_4$ nanosheets was formed (Figure 6.5b), which increases the specific surface area of the photocatalyst and the separation of the photogenerated charge carriers was enhanced (Zhong et al. 2018). The photocatalytic hydrogen production mechanism of the C-TiO$_2$/g-C$_3$N$_4$ is shown in Figure 6.5c. Due to the narrow band gap of C-TiO$_2$ (2.81 eV) and g-C$_3$N$_4$ (2.94 eV), photoinduced electrons and holes can be generated on C-TiO$_2$ and g-C$_3$N$_4$ under visible light. The electrons generated in the CB of g-C$_3$N$_4$ can rapidly move to the CB of C-TiO$_2$, while the holes transferred from the VB of C-TiO$_2$ to the VB of g-C$_3$N$_4$. Therefore, the heterojunction of C-TiO$_2$ and g-C$_3$N$_4$ can effectively accelerate the transmission of the charge carriers and avoid the recombination of photogenerated holes and electrons (Yuan et al. 2020), so that more photogenerated electrons can participate in the photocatalytic hydrogen evolution reaction.

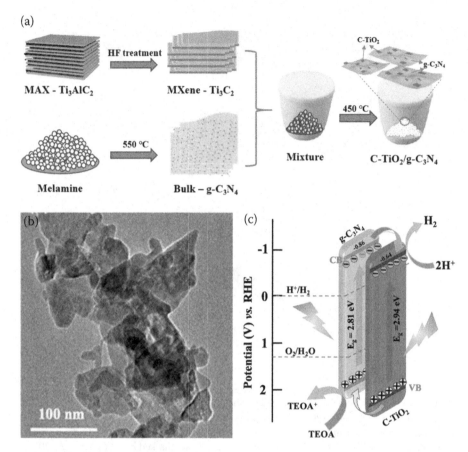

FIGURE 6.5 (a) Schematic illustration of the C-TiO$_2$/g-C$_3$N$_4$ photocatalysts preparation process. (b) TEM image of C-TiO$_2$/g-C$_3$N$_4$. (c) Schematic diagram of the photocatalytic hydrogen production mechanism for C-TiO$_2$/g-C$_3$N$_4$ photocatalysts (Han et al. 2020).

6.2.3 SEMICONDUCTOR/MXENE COMPOSITE PHOTOCATALYST

MXene, with excellent conductivity and suitable Fermi energy level, can be used as a co-catalyst for photocatalytic hydrogen evolution. Semiconductor/MXene composite photocatalysts can be prepared by in situ growth method (Xiao et al. 2020), physical mixing (Liao et al. 2020), and electrostatic self-assembly (Zhuang, Liu, and Meng 2019), et. al. For the semiconductor/MXene system, the tight heterojunction formed between the semiconductor and MXene can act as the charge carriers transfer channel to facilitate the transfer of photogenerated electrons and holes (Zhao et al. 2019) and enhanced the photocatalytic performance. The photocatalytic hydrogen evolution rates of semiconductor with different co-catalysts were shown in the Table 6.2, indicating that Ti$_3$C$_2$ as a co-catalyst can meet the requirements of increasing photocatalytic activity like other co-catalysts.

Zn$_2$In$_2$S$_5$, as typical ternary sulfide, was one of the most popular photocatalysts in the field of photocatalytic research because of its unique electronic properties and

TABLE 6.2

Comparison of MXene-Based Photocatalysts with Other Photocatalysts

Photocatalyst	Cocatalyst	Amount of Catalyst (mg)	Sacrificial Reagent	Light Condition	H_2 Production Rate ($\mu mol \cdot g^{-1} \cdot h^{-1}$)	Multiple of Enhance*	References
$Zn_2In_2S_5/Ti_3C_2$	Ti_3C_2	50	0.25 M Na_2SO_3, 0.35 M Na_2S	300 W Xe lamp (420 nm cutoff filter)	12938.80	1.97 times ($Zn_2In_2S_5$)	(Wang, Sun, et al. 2019)
$ZnIn_2S_4/MoS_2$	MoS_2	/	10 vol% lactic acid	300 W Xe lamp (420 nm cutoff filter)	4974.00	50 times ($ZnIn_2S_4$)	(Huang et al. 2019)
$ZnIn_2S_4$	/	3	/	300 W Xe lamp (420 nm cutoff filter)	1940.00	11.2 times ($ZnIn_2S_4$)	(Shi et al. 2020)
$Cu_x-ZnIn_2S_4$	Cu, Pt	100	20 vol% ascorbic acid	500 W Xe lamp (AM 1.5 G filter)	26200.00	4.0 times ($ZnIn_2S_4$)	(Wang, Shen, et al. 2019)
$Ti_3C_2T_x/TiO_2$	$Ti_3C_2T_x$	30	25 vol% methanol	200 W Hg lamp (285–325 nm cutoff filter)	920.00	9.1 times (TiO_2)	(Su et al. 2019b)
Ti_3C_2/TiO_2	Ti_3C_2	50	10 vol% glycerin	350 W Xe lamp	120.00	/	(Li, Zhang, et al. 2019)
$Cu/TiO_2@Ti_3C_2T_x$	$Ti_3C_2T_x$, Cu	20	6 vol% methanol	300 W Xe lamp	860.00	10.0 times ($TiO_2@Ti_3C_2Tx$)	(Peng et al. 2018)
TiO_2/Ti_3C_2	Ti_3C_2	50	10 vol% methanol	300 W Xe lamp	6979.00	3.8 times (TiO_2)	(Zhuang, Liu, and Meng 2019)
$MoS_2/Ti^{3+}-TiO_2$	MoS_2	50	10 vol% TEOA	300 W Xe lamp	713.15	15.0 times (TiO_2)	(Ou et al. 2020)
$TiO_2-Au-CdS$	Au	50	0.1 M Na_2S, 0.1 M Na_2SO_3	300 W Xe lamp (420 nm cutoff filter)	572.30	3.1 times (TiO_2-CdS)	(Yao et al. 2020)
$Ti_3C_2/g-C_3N_4$	Ti_3C_2, Pt	30	10 vol% TEOA	200 W Hg lamp (400 nm cutoff filter)	72.30	10.0 times ($g-C_3N_4$)	(Su et al. 2019a)

(Continued)

TABLE 6.2 (Continued)
Comparison of MXene-Based Photocatalysts with Other Photocatalysts

Photocatalyst	Cocatalyst	Amount of Catalyst (mg)	Sacrificial Reagent	Light Condition	H_2 Production Rate ($\mu mol \cdot g^{-1} \cdot h^{-1}$)	Multiple of Enhance*	References
g-C$_3$N$_4$/Ti$_3$C	Ti$_3$C$_2$	50	10 vol% TEOA	300 W Xe lamp (420 nm cutoff filter)	116.20	6.6 times (g-C$_3$N$_4$)	(Li, Zhao, et al. 2020)
g-C$_3$N$_4$/Ti$_3$C$_2$	Ti$_3$C$_2$, Pt	30	10 vol% TEOA	300 W Xe lamp	5100.00	5.0 times (g-C$_3$N$_4$)	(An et al. 2018)
Co-S-X/g-C$_3$N$_4$	Co-S-X, Pt	5	12 vol% TEOA	300 W Xe lamp (AM 1.5 G cutoff filter)	10328.90	396.0 times (g-C$_3$N$_4$)	(Chen, Xu, et al. 2020)
MoS$_2$/g-C$_3$N$_4$	MoS$_2$	20	20 vol% TEOA	300 W Xe lamp (AM 1.5 G cutoff filter)	1124.00	5.9 times (g-C$_3$N$_4$)	(Sun, Yang, et al. 2020)
g-C$_3$N$_4$/GDY	graphdiyne	20	15 vol% TEOA	300 W Xe lamp (420 nm cutoff filter)	454.28	3.0 times (g-C$_3$N$_4$)	(Si et al. 2020)

* The ratio of the photocatalytic hydrogen production rate of the composite to that of the pure photocatalyst (in brackets).

suitable band structure (Li, Cheng, et al. 2019; Wang, Sun, et al. 2019). The two-dimensional layered Ti_3C_2 was used as a platform for the generation of $Zn_2In_5S_5$ nanosheets microspheres in situ on its surface by a simple hydrothermal reaction (Figure 6.6a) (Wang, Sun, et al. 2019). From the differential charge density in $Zn_2In_2S_5/Ti_3C_2$ shown in Figure 6.6b, it can be seen that the surface metal atoms of Ti_3C_2 extract electrons from the metal atoms of the neighboring $Zn_2In_2S_5$ at the interface. In addition, an internal electric field was formed due to the interfacial-built-in quasi-alloying effect, which was beneficial to the transfer and separation of the photogenerated electrons and holes (Hu et al. 2016). In this $Zn_2In_2S_5/Ti_3C_2$ system, the photogenerated electrons were excited from the S 2p orbital in $Zn_2In_2S_5$ to the In 5s5p orbital, and then rapidly injected into the Ti 3d orbital in Ti_3C_2. Besides, the H_3O^+ ions were adsorbed on the active sites located in the middle layer or edge of Ti_3C_2 and combined with photoinduced electrons to form H* atoms. Subsequently, the H* atom reacts with another H_3O^+ ion and the electrons from

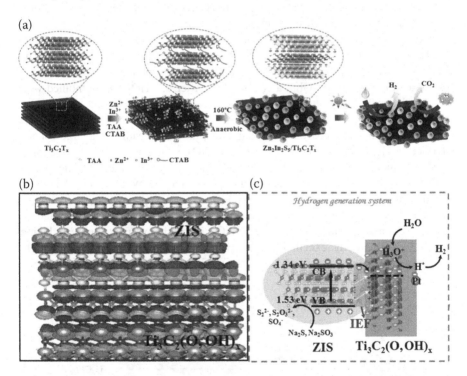

FIGURE 6.6 (a) Schematic illustration of the synthesis process of $Zn_2In_2S_5/Ti_3C_2$ (O, OH)$_x$. TAA = thioacetamide, CTAB = cetyltrimethylammonium bromide. (b) The differential charge density of $Zn_2In_2S_5/Ti_3C_2$. The differential charge density is defined as the difference in electronic density before and after bonding, in which the S atoms, In atoms, Zn atoms, C atoms, and O atoms were marked in yellow, purple, silver, light blue, brown, and red, respectively. The spheres in green and dark blue represented positive and negative charges, respectively. (c) Schematic of the mechanism for photocatalytic hydrogen generation. ZIS = $ZnIn_2S_4$, CB = conduction band, VB = valence band, IEF = interface (Wang, Sun, et al. 2019).

Ti$_3$C$_2$ to form the H$_2$ molecule. With the gradual accumulation of hydrogen atoms, the hydrogen atoms on Ti$_3$C$_2$ reacted to form the hydrogen, and the hydrogen was released from the surface (Sun et al. 2018), as shown in Figure 6.6c. Stimulating by visible light, photogenerated electron-hole pairs can be produced in Zn$_2$In$_2$S$_5$. Since the Fermi level of Ti$_3$C$_2$ was more positive than the conduction band of Zn$_2$In$_2$S$_5$, the photogenerated electrons transferred from the Zn$_2$In$_2$S$_5$ to Ti$_3$C$_2$ through the intimate interface. Under the action of quasi-metallization, the weak bond between the interface and the positively charged species (H$_3$O$^+$ and H*) facilitates the rapid dissociation and precipitation of hydrogen (Greiner et al. 2018).

In addition to the multilayer morphology with the accor,dion structure of MXene, monolayer MXene nanosheets can be prepared by etching and ultrasonic processing (Naguib et al. 2014). Compared with multilayer Ti$_3$C$_2$, monolayer Ti$_3$C$_2$ not only had larger specific surface area and more exposed active sites, but also the shortened transfer distance of photogenerated charge carriers for the photocatalytic reaction. Monolayer co-catalysts can effectively enhance the separation of photo-generated electrons and holes (Du et al. 2019; Jiang et al. 2020).

Ti$_3$C$_2$T$_x$/TiO$_2$ was prepared by physical mixing of monolayer Ti$_3$C$_2$T$_x$ and TiO$_2$ nanoparticles under the assisted ultrasonic treatment (Su et al. 2019b). The monolayer Ti$_3$C$_2$T$_x$. Ultrathin Ti$_3$C$_2$T$_x$ nanosheets were prepared, with the thickness of about 1.7–2.2 nm (Ghidiu et al. 2014). After the TiO$_2$ was combined with the monolayer Ti$_3$C$_2$T$_x$, the specific surface area was enhanced and more active sites were exposed. Moreover, the light-absorption capacity of the composites was increased, which can improve the utilization of sunlight. In addition, a large contact interface can be formed between Ti$_3$C$_2$T$_x$ and TiO$_2$, which is beneficial to the rapid transfer of the photogenerated electrons and holes. Therefore, the TiO$_2$/monolayer Ti$_3$C$_2$T$_x$ composites showed a much higher photocatalytic hydrogen evolution rate than that of TiO$_2$/multilayer Ti$_3$C$_2$T$_x$ composites. For the photocatalytic system of TiO$_2$/monolayer Ti$_3$C$_2$T$_x$, the photoinduced electrons and holes were generated on the conduction band and valence band of TiO$_2$, respectively, under light irradiation (Figure 6.7). Due to the excellent conductivity of monolayer Ti$_3$C$_2$T$_x$ and the tight contact between monolayer Ti$_3$C$_2$T$_x$ and TiO$_2$, photogenerated electrons can rapidly

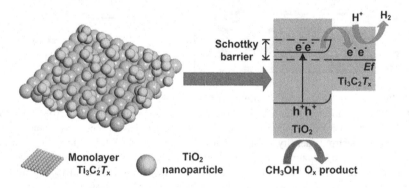

FIGURE 6.7 Schematic illustration of the mechanism for photocatalytic hydrogen evolution over the Ti$_3$C$_2$T$_x$/TiO$_2$. E$_F$ = Fermi level (Su et al. 2019b).

transfer from CB of TiO_2 to $Ti_3C_2T_x$, and the electrons on the surface of the $Ti_3C_2T_x$ could further participate in the photocatalytic-reduction reaction (Chen, Huang, et al. 2021) (Figure 6.8).

Electrostatic self-assembly is a strategy to prepare composite by the strong electrostatic effect caused by surface charge (Shi, Liu, and He 2018; Fu et al. 2019; Chen, He, et al. 2020). For example, the Ti_3C_2 displayed a negative surface due to the negative functional group (F-, O-, OH-) on its surface (Guo et al. 2019; Zhang et al. 2020), and the semiconductor/Ti_3C_2 can be prepared by electrostatic self-assembly with the protonated semiconductor. Graphitic carbon nitride (g-C_3N_4) was a nonmetallic visible-light response photocatalyst, and the ultrathin g-C_3N_4 nanosheets can be obtained by thermal polymerization (Wang, Hong, et al. 2020). When the monolayer Ti_3C_2 was coupled with g-C_3N_4 nanosheet, a tight 2D/2D heterojunction can form, which is conducive to the charge transfer between the

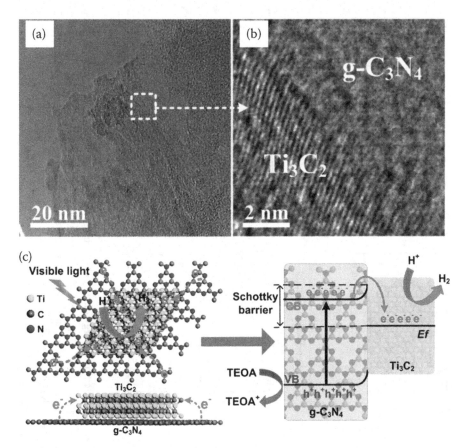

FIGURE 6.8 TEM (a) and HRTEM (b) TEM images of Ti_3C_2/g-C_3N_4. The TEM image pointed by the white arrow was an enlargement of the selected area of the white dotted box. (c) Schematic illustration of the mechanism for photocatalytic hydrogen evolution over the $Ti_3C_2T_x$/TiO_2. E_F = Fermi level, CB = conduction band, VB = valence band. TEOA = triethanolamine (Su et al. 2019a).

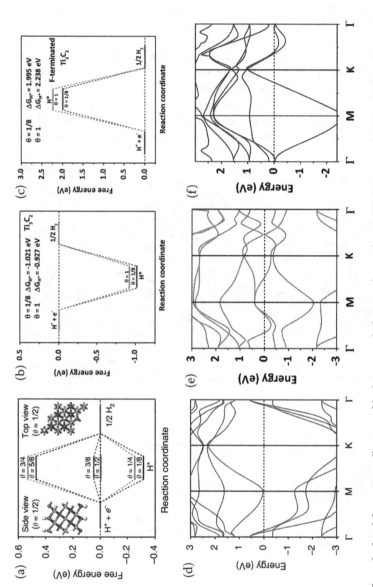

FIGURE 6.9 The calculated free-energy diagram of hydrogen evolution reaction at the equilibrium potential (U = 0 V) on the surface of a 2 × 2 × 1 (a) O-terminated Ti_3C_2 (the side and top views of a 2 × 2 × 1 O-terminated Ti_3C_2 supercell at 1/2 H* coverage was shown in the inset), (b) bare Ti_3C_2, and (c) F-terminated Ti_3C_2 supercell at different H* coverage. The calculated band structure of (d) O-terminated Ti_3C_2, (e) bare Ti_3C_2, and (f) F-terminated Ti_3C_2 (Ran et al. 2017).

Ti_3C_2 and g-C_3N_4 (Su et al. 2019a). The 2D/2D heterojunction had obvious advantages over 0D/2D (nanoparticles/monolayer Ti_3C_2) heterojunction due to its large contact interface (Li, Wang, et al. 2019). Moreover, the 2D/2D heterojunction shortened the transfer distance and time of photogenerated electrons to the photocatalyst surface (Dong et al. 2020). The thickness of the $Ti_3C_2T_x$ nanosheets was about 1.7 nm, and the 2D $Ti_3C_2T_x$ structure consists of nanosheets with irregular shapes. As shown in Figures 6.9a and b, the clear lattice fringes were designated as Ti_3C_2 nanosheets, while the disordered regions were attributed to g-C_3N_4 nanosheets. The mechanism for photocatalytic hydrogen evolution over 2D/2D $Ti_3C_2T_x$/g-C_3N_4 composites was shown in Figure 6.9c. Under light irradiation, the $Ti_3C_2T_x$ can effectively attract the photoinduced electrons from the conduction band of g-C_3N_4 due to the 2D/2D tight heterojunctions, thus effectively avoiding the recombination of the photogenerated electrons and holes and enhanced the photocatalytic-hydrogen evolution.

6.3 MECHANISM OF MXENE-BASED PHOTOCATALYST FOR PHOTOCATALYTIC WATER-SPLITTING

6.3.1 ROLE OF MXENE WITH DIFFERENT TERMINATING GROUPS IN PHOTOCATALYTIC WATER-SPLITTING

The surface-termination group had a great influence on the physical and chemical property of MXene (Liu, Ding, et al. 2020). For Ti_3C_2, the surface-termination groups were mainly including the F, OH, and O (Rosenkranz et al. 2019). The difference in termination groups on the surface generally depends on the preparation method of MXene. When HF solution was selected as the etching agent, the termination group on Ti_3C_2 was mainly OH. In contrast, when HCl/LiF solution was selected as the etching agent, surface groups were mainly O. In addition to the effect of preparation condition, the surface-termination group of Ti_3C_2 can also be changed by post-treatment. The F groups can gradually be replaced by the O termination in the air, and the OH groups can be converted to the O groups at 200 °C. The O-termination was more stable than F-termination and OH-termination on MXene.

MXene was mainly used as a co-catalyst in the field of photocatalytic water-splitting. An efficient co-catalyst can attract the photogenerated electrons from the photocatalyst to its surface. Then, H_2 can produce on the co-catalyst surface with high efficiency. Different termination groups of $Ti_3C_2T_x$ exhibit different charge transfer rates and active sites. The DFT calculation of bare Ti_3C_2, F-terminated Ti_3C_2, and O-terminated Ti_3C_2 demonstrated that the Ti_3C_2 with difference termination shows different photocatalytic hydrogen-production performance (Ran et al. 2017). Photocatalytic hydrogen production can be divided into three stages: (I) initial state $H^+ + e^-$, (II) intermediate adsorbed H*, and (III) final product 1/2 H_2. The Gibbs free energy of the intermediate state, $|\Delta G_{H*}|$, was regarded as an indicator of the hydrogen-evolution reaction activity. The photocatalytic hydrogen-evolution performance is of high efficiency when the $|\Delta G_{H*}|$ approached to zero (Jiao et al. 2015; Pandey and Thygesen 2017). As shown in Figures 6.9a, b, and c, in the optimum H* coverage (θ = 1/2), the $|\Delta G_{H*}|$ of O-terminated Ti_3C_2 equal to 0.00283 eV, close to zero value, indicating that O-terminated Ti_3C_2 has excellent hydrogen-evolution activity. However, bare Ti_3C_2 shows a large negative Gibbs free energy (−0.927 eV), indicating that the chemical adsorption of H* on the surface was too strong, which was not conducive to the desorption of the product. F-terminated Ti_3C_2 shows a positive Gibbs free energy (1.995 eV), indicating that chemical adsorption of H* on the surface was too weak. Therefore, bare Ti_3C_2 and F-terminated Ti_3C_2 were not good at photocatalytic hydrogen-evolution reaction.

An ideal co-catalyst can draw the photoproduction electrons efficiently from photocatalyst and transfer the electrons to its surface quickly, which requires appropriate band structure and excellent electrical conductivity. The band structure of O-terminated Ti_3C_2, bare Ti_3C_2, and F-terminated were shown in Figures 6.9d, e, and f. Unlike the semiconductors, Ti_3C_2 did not have a band gap, so it can only be used as a co-catalyst in the photocatalytic reaction. Bare Ti_3C_2 exhibits metallic property with electronic states spanning the Fermi level. The continuous-crossing Fermi level suggest that the conductivity of O-terminated Ti_3C_2 and F-terminated Ti_3C_2 was not as good as the bare Ti_3C_2, but their electrical conductivity is still excellent. Meanwhile, Fermi level (E_F) was an important index to study the mechanism of MXene for photocatalytic hydrogen production. Fermi level (E_F) of O-terminated Ti_3C_2, bare Ti_3C_2, and F-terminated Ti_3C_2 was 1.88 V, −0.05 V, and 0.15 V, respectively, versus standard hydrogen electrode (SHE). The more positive of the Fermi level, the easier it was to capture electrons from the conduction band (CB) of the semiconductor. The O-terminated Ti_3C_2 showed the most positive value of Fermi level (E_F), which indicated that the ability to attract electrons of O-terminated Ti_3C_2 is better than that of bare Ti_3C_2 and F-terminated Ti_3C_2.

For the 2D/2D MXene/Bi_2WO_6 photocatalytic system (Cao et al. 2018), the Fermi energy level of Ti_3C_2 with different terminals are shown in Figure 6.10. According to Equations 6.1 and 6.2, the Fermi level (E_F) was converted to the normal hydrogen electrode (NHE) and vacuum level.

$$E_F(vs\ \text{NHE},\ \ \text{pH} = 7) = E_F(vs\ \text{NHE},\ \ \text{pH} = 0) - 0.59\ \text{pH} \qquad (6.1)$$

$$E_F(vs\ \text{NHE},\ \ \text{pH} = 0) = -4.5\ V - E_F(vs\ \text{vacuum level}) \qquad (6.2)$$

The transfer path of photogenerated electrons and holes can be studied by calculating the work function. The electron transfers from a position where the work function is low to a position where the work function is high. Instead, the hole transfers from the position where the work function is high to the position where the work function is low. The Fermi level (E_F) of bare Ti_3C_2, F-terminated Ti_3C_2, and O-terminated Ti_3C_2 was −0.45 V, 0.18 V, and 0.71 V (vs NHE, pH = 7). The O-terminated Ti_3C_2 shows the most positive Fermi level (E_F), proving a strong attraction to electrons in the semiconductor conduction band (CB). The work function of bare Ti_3C_2, F-terminated Ti_3C_2, and O-terminated Ti_3C_2 was 4.46 eV, 5.09 eV, and 5.62 eV, which indicates that the O-terminated Ti_3C_2 with higher work function had a stronger ability to capture electrons (Liu, Xiao, and Goddard 2016).

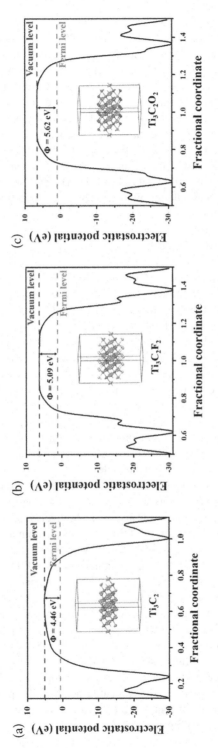

FIGURE 6.10 The Fermi level (E_F) of (a) bare Ti_3C_2, (c) F-terminated Ti_3C_2, and (c) O-terminated Ti_3C_2 (Cao et al. 2018).

In addition to O-terminated Ti_3C_2 and F-terminated Ti_3C_2, OH-terminated Ti_3C_2 was also applied for photocatalysis (Peng et al. 2018). For example, (001) TiO_2/ Ti_3C_2 composite with (001) TiO_2 nanosheets generated in situ on Ti_3C_2 was prepared by hydrothermal oxidation (Peng et al. 2016). According to the results of FTIR and XPS, Ti_3C_2 was speculated to be the OH-terminated Ti_3C_2. OH-terminated Ti_3C_2 shown an ultra-low work function of about 1.8 eV, which was much lower than that of (001) TiO_2 (4.9 eV). The Schottky barriers were formed at interfaces of TiO_2 and Ti_3C_2. Since the work function of OH-terminated Ti_3C_2 was lower than that of (001) TiO_2, the photogenerated holes in the valence band (VB) of TiO_2 were transferred to Ti_3C_2. Interestingly, a downward curvy Schottky junction can be formed between the (001) TiO_2/Ti_3C_2 interfaces. The synergistic effect of the Schottky junction and OH-terminated Ti_3C_2 can effectively capture photogenerated holes and avoid the recombination of the photogenerated electrons and holes on TiO_2.

MXene/semiconductor photocatalyst can be synthesized using MXene as a template. The semiconductor with corresponding elements was prepared from the atoms exposed on MXene as the source. Notably, MXene with different terminations had an impact on this process. For instance, researchers studied the photocatalytic hydrogen-evolution activity of TF (TiO_2/Ti_3C_2 obtained by calcination of F-terminated Ti_3C_2) and TOH (TiO_2/Ti_3C_2 obtained by calcination of OH-terminated Ti_3C_2) (Li, Zhang, et al. 2019). The preparation process can be simply divided into two steps. In step I, Ti_3C_2 was treated with deionized water and NaOH respectively. Characterization showed that the two solutions did not change the crystal structure and chemical composition of Ti_3C_2. In step II, F-terminated Ti_3C_2 and OH-terminated Ti_3C_2 products were calcined and marked as TF and TOH. For the TF sample, the residual Ti_3C_2 still presented a layered structure, and TiO_2 generated in situ at the edge of the nanosheets with a truncated octahedral bipyramid structure (Li, Zhang, et al. 2019). In contrast, TiO_2 nanoparticles were generated in situ at the edge of the TOH sample.

The truncated octahedral bipyramid TiO_2 exposed more (001) facets, which form a heterojunction with (001) facets of TiO_2. The formation of heterojunction was advantageous to the separation and transfer of photogenerated charge. The photocatalytic hydrogen production rate of TF was two times higher than that of TOH, which indicated that the exposed (001) facets were a key factor for improving the photocatalytic activity. Under the light irradiation, the (001) and (101) facets of TiO_2 can be excited to generate electrons and holes. Because the atoms were arranged differently on the surface of each facet, the band gap of the (001) facets were not aligned with the band gap of the (101) facets, indicating the formation of heterojunction on the surface of (001) facets and (101) facets. Photogenerated electrons can rapidly transfer from the (001) facet to the (101) facet, avoiding the recombination of photogenerated electrons and holes. Residual Ti_3C_2 relied on its excellent conductivity can obtain electrons from the conduction band of TiO_2 (101) facets for the reduction reaction. The truncated octahedral bipyramid TiO_2 can be obtained by calcination of the F-terminated Ti_3C_2. Due to the exposed (101) and (001) of truncated octahedral bipyramid TiO_2, the heterojunction was constructed to effectively promote the transfer and separation of photogenerated electron-hole pairs, and the photocatalytic hydrogen-production rate was greatly improved (Meng et al. 2016; Liu et al. 2017).

6.3.2 SEMICONDUCTOR/MXENE PHOTOCATALYTIC SYSTEM WITH SCHOTTKY JUNCTION

For $Ti_3C_2T_x$ with different termination groups, the F-termination and OH-termination can be easily replaced by the O-termination (Sun et al. 2018), and the O-termination on the surface and edge of Ti_3C_2 act as the active site for hydrogen-evolution reaction (marked with *). The reaction mechanism of hydrogen evolution on the surface of Ti_3C_2 including three main steps (Deng et al. 2014; Ling et al. 2016).

$$H_3O^+ + e^- + * \rightarrow H^* + H_2O \,(\text{Volmer reaction}) \tag{6.3}$$

$$H_3O^+ + e^- + H^* \rightarrow H_2 + H_2O \,(\text{Heyrovsky reaction}) \tag{6.4}$$

$$H^* + H^* \rightarrow H_2 \,(\text{Tafel reaction}) \tag{6.5}$$

The Volmer reaction mechanism is shown in (6.3). H_3O^+ ions can be adsorbed at the Ti_3C_2 active site (O-termination) and then combined with electrons to form H atoms (He et al. 2018). At the same active site (O-termination), the H atom combined with another H_3O^+ ion from water and an electron from Ti_3C_2 to form the H_2 molecule, which is called the Heyrovsky reaction mechanism, as shown in (6.4). Finally, for the Tafel reaction mechanism (6.5), H_2 molecule can be formed by two H atoms at the same active site (Bhardwaj and Balasubramaniam 2008). In addition to providing the active sites, the Schottky junction can also be formed by the suitable Fermi level of Ti_3C_2 and semiconductors, which is beneficial to the separation of photoinduced electrons and holes.

The suitable Fermi level of Ti_3C_2 enabled it to form heterojunction at the interface with the semiconductor, as shown in Figure 6.11. In the heterojunction, the semiconductor lost electrons and became positively charged, and the Ti_3C_2 accepted the electrons to form the space-charge layer at the interface. In this case, the conduction band (CB) and valence band (VB) of semiconductor bend upward due to the existence of the space-charge layer, which led to the formation of the Schottky junction. Under light irradiation, the electrons on the VB of the semiconductor excited into the CB, and the upward curved CB allowed electrons to accumulate in the space-charge layer. When the energy of the accumulated electron exceeds the Schottky barrier, the electrons can transfer across the curved CB to MXene and quickly transfer to the active site on the surface due to the excellent electronic conductivity of MXene.

Sandwich-like layered heterostructure (UZNs-MNs-UZNs) was prepared by using the $Ti_3C_2T_x$ as the support for in situ growth of ultrathin $ZnIn_2S_4$ nanosheet (UZNs) on its surface, and it was used for photocatalytic hydrogen evolution from water. In this system, the surface-metal atoms (Ti) of Ti_3C_2 accept electrons from the metal atoms (Zn) of $ZnIn_2S_4$ via the interface. As the migration of electrons, $ZnIn_2S_4$ shows the positive charges, which lead to the formation of a space-charge

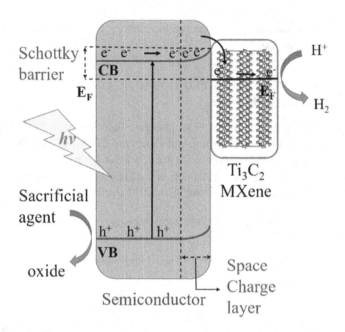

FIGURE 6.11 Schematic of photocatalytic hydrogen evolution mechanism for semi-conductor/Ti_3C_2 MXene photocatalyst Schottky junction system. E_F = Fermi level, CB = conduction band, VB = valence band, e^- = electrons, h^+ = holes.

layer at the interface. The presence of a space-charge layer causes the CB of $ZnIn_2S_4$ to bend upward. The Fermi level potential of the $Ti_3C_2T_x$ was more positive than the CB potential of $ZnIn_2S_4$, which leads to the formation of the Schottky junction. When the electrons can overcome the Schottky barrier, the electrons can migrate to the Fermi level of Ti_3C_2 and then quickly move to the surface of Ti_3C_2 for the reduction reaction.

6.3.3 MXene Acts as a Bridge for the Separation of Photogenerated Charge Carriers

As for the photocatalytic hydrogen-production mechanism of MXene-based photocatalysis, Ti_3C_2 MXene can not only be used to form the Schottky junction and act as the active site of photocatalytic hydrogen production, but it can also act as a bridge for the separation of photogenerated charge carriers in photocatalytic hydrogen-production reaction. For example, a $Ti_3C_2(TiO_2)@CdS/MoS_2$ composite photocatalyst was designed and prepared systematically, and the role of Ti_3C_2 is to act as a bridge rather than an active site for photocatalytic hydrogen production (Ai et al. 2019). As shown in Figure 6.12, the band structure of $Ti_3C_2(TiO_2)@CdS/MoS_2$ realized the separation of photogenerated electrons and holes. The photogenerated electrons can transfer from the CB of CdS to the CB of MoS_2, realizing the separation of electron-hole pairs. Meanwhile, Ti_3C_2 serves as the mediator for attracting holes because of its oxidation feature (Enyashin and Ivanovskii 2012).

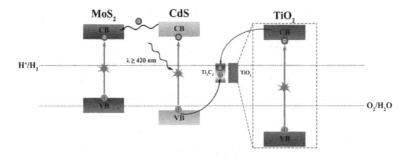

FIGURE 6.12 Schematic illustration of the band structure of $Ti_3C_2(TiO_2)@CdS/MoS_2$ system (Ai et al. 2019).

Thus, the photogenerated holes of CdS move to Ti_3C_2, which reduces the probability of recombination. Moreover, the Z-scheme between CdS and TiO_2 further enhanced the separation of electron-hole pairs and greatly prolonging the lifetime of photogenerated electrons.

Ultrathin g-C_3N_4 (UCN) and black phosphorus (BP) quantum dots (BQ) were prepared separately and loaded on different faces of ultrathin Ti_3C_2 nanosheets to construct the ternary BQ/TiC/UCN heterojunction photocatalyst. The incorporation of Ti_3C_2 nanosheets substantially accelerates the transfer and the separation of photogenerated charge carriers between BP quantum dots and ultrathin g-C_3N_4. The type II heterojunction was formed at the BQ/TiC/UCN composites with Ti_3C_2 as the bridge. The implanted Ti_3C_2 nanosheets can form compact interfaces with BP quantum dots and ultrathin g-C_3N_4, respectively. Therefore, the electrons in the CB of BP quantum dots can quickly migrate to the Ti_3C_2, and then the electrons rapidly migrate to the CB of ultrathin g-C_3N_4. As a result, the photogenerated electrons can effectively react with protons to produce the H_2. Correspondingly, the holes on the VB of ultrathin g-C_3N_4 passed through the Ti_3C_2 to the VB of BP quantum dots for oxidation reaction.

6.4 STABILITY OF MXENE FOR THE PHOTOCATAYTIC WATER-SPLITTING

Stability of the photocatalyst is very important for the practical application (Ma et al. 2020). Ti_3C_2 MXene, especially monolayer Ti_3C_2, was not stable in water and at high temperature (He et al. 2020). Due to the ultrathin two-dimensional structure of the Ti_3C_2, the titanium atoms on the edge and surface can be oxidized to TiO_2 (Xu, Wang, et al. 2018). However, Ti_3C_2 exhibited excellent stability in the photocatalytic hydrogen-production reaction. When MXene was used as co-catalyst in the photocatalytic hydrogen-production reaction, the photogenerated electrons were transferred to Ti_3C_2 rather than the holes to Ti_3C_2, which can avoid the oxidation of Ti_3C_2. Therefore, the oxidation of Ti_3C_2 was mainly caused by the dissolved oxygen in the reaction solution.

$Ti_3C_2(TiO_2)@CdS/MoS_2$ quaternary composite photocatalyst was prepared and applied for the photocatalytic hydrogen production (Ai et al. 2019). The photocatalytic activity was investigated in sacrificial agent solution (8.47 mmol·h^{-1}·g^{-1})

and pure water ($344.74\ \mu mol\cdot h^{-1}\cdot g^{-1}$). The photocatalytic hydrogen-production activity was maintained after reaction for 100 h, which preserved the original hydrogen production rate of 98% in lactic acid solution. For the Ti_3C_2 (TiO_2)@CdS/MoS_2 photocatalytic system, the Ti_3C_2 nanosheets were slowly oxidized in water. In the oxidation process, a TiO_2 layer was gradually formed on the surface of Ti_3C_2 nanosheets, which could be used as a shielding layer to delay the oxidation of Ti_3C_2 nanosheets and account for the excellent stability of the photocatalyst.

The stability of Ti_3C_2 nanosheet can also be enhanced by Al^{3+} intercalation (Ding et al. 2020). The Ti_3AlC_2 had excellent stability at high temperature and in aqueous solution. However, the Ti_3C_2 nanosheets obtained from the etching Ti_3AlC_2 were difficult to maintain the stability. The reason was that there was a stable Ti-Al metallic bond in the MAX phase, which was removed by etching out the Al layer. Through Al^{3+} intercalation, the $Ti_3C_2T_x$-Al^{3+} ionic bond was formed with oxygen-containing groups on the surface of Ti_3C_2, with the restult being an enhanced stability of the 2D Ti_3C_2 nanosheets.

6.5 CONCLUSION AND PROSPECTS

In summary, this chapter summarizes the design, preparation, and mechanism of MXene-based photocatalysts for the photocatalytic hydrogen evolution in recent years. The excellent conductivity and two-dimensional structure of MXene can effectively promote the separation of electron holes, thus enhancing the photocatalytic hydrogen-evolution activity. Therefore, MXene has been widely used as an efficient co-catalyst for photocatalytic hydrogen-evolution reactions. Different etching and stripping methods of MXene and preparation methods of different MXene-based catalysts provide different ideas for the application of MXene in the field of photocatalysis. The study of different surface-termination groups of MXene provides a basis for exploring the mechanism of photogenerated electron transfer and photocatalysis. The research on the stability of MXene-based photocatalysts is focused on overcoming the limitation of the practical and industrial application of photocatalytic hydrogen production.

The variable etching environment provides easy control options for the preparation of MXene and offers new opportunities to improve the synthesis process. However, the selection of HF for etching processes can cause environmental pollution, so the study of environmentally friendly methods for large-scale synthesis of MXenes may be an attractive and urgently needed area. Meanwhile, the metal properties, surface group arrangement, and high conductivity of 2D $Ti_3C_2T_x$ are important factors to improve photocatalytic activity. However, in future studies, the stability and complexity of the 2D $Ti_3C_2T_x$ MXene photocatalytic process need to be understood. In addition, the research work needs to be extended to one and three-dimensional MXene, and the unknown characteristics of MXene-based photocatalytic systems with different morphologies still need to be studied and discovered.

Furthermore, an insightful understanding of the mechanism of photocatalysis in the photocatalytic process remains a huge challenge. To design MXene-based photocatalysts to improve their photocatalytic activity, the photocatalytic mechanism needs to be determined, which means that the relationship between the synthesis-reaction process and the physical and chemical properties of reactant products, as well as the factors affecting the photocatalytic efficiency, needs to be understood. In the field of photocatalytic hydrogen production, the study on the adsorption and desorption behavior of reactants and products on the surface of MXene and the active site is the key to improve the MXene-based photocatalyst. In this regard, many advanced research methods need to be developed for research and analysis. Finally, the oxidation rate of MXene in water is high, and the physical and chemical properties would change significantly; the stability of MXene is the key to the industrialization of MXene photocatalysis. In addition, in previous studies, MXene was often used as a cocatalyst, and MXene alone as a photocatalyst can carry out photocatalytic reactions more efficiently, which will be a research focus in the future.

ACKNOWLEDGMENTS

This work was supported by the National Natural Science Foundation of China (22078074, 21938001), Guangxi Natural Science Foundation (2019GXNSFAA-245006, 2020GXNSFDA297007, 2016GXNSFFA380015), Special funding for 'Guangxi Bagui Scholars', and Scientific Research Foundation for High-level Personnel from Guangxi University.

REFERENCES

Ai, Zizheng, Yongliang Shao, Bin Chang, Baibiao Huang, Yongzhong Wu, and Xiaopeng Hao. 2019. Effective orientation control of photogenerated carrier separation via rational design of a $Ti_3C_2(TiO_2)@CdS/MoS_2$ photocatalytic system. *Applied Catalysis B: Environmental* 242:202–208.

Ai, Zizheng, Kang Zhang, Bin Chang, et al. 2020. Construction of $CdS@Ti_3C_2@CoO$ hierarchical tandem p-n heterojunction for boosting photocatalytic hydrogen production in pure water. *Chemical Engineering Journal* 383:123130.

Alhabeb, Mohamed, Kathleen Maleski, Babak Anasori, et al. 2017. Guidelines for synthesis and processing of two-dimensional titanium carbide (Ti_3C_2Tx MXene). *Chemistry of Materials* 29 (18):7633–7644.

An, Xiaoqiang, Wei Wang, Jiangpeng Wang, Haozhi Duan, Jintao Shi, and Xuelian Yu. 2018. The synergetic effects of Ti_3C_2 MXene and Pt as co-catalysts for highly efficient photocatalytic hydrogen evolution over g-C_3N_4. *Physical Chemistry Chemical Physics* 20 (16):11405–11411.

Arabi Shamsabadi, Ahmad, Mohammad Sharifian Gh, Babak Anasori, and Masoud Soroush. 2018. Antimicrobial mode-of-action of colloidal Ti_3C_2Tx Mxene nanosheets. *ACS Sustainable Chemistry & Engineering* 6 (12):16586–16596.

Bhardwaj, Mukesh and R. Balasubramaniam. 2008. Uncoupled non-linear equations method for determining kinetic parameters in case of hydrogen evolution reaction following Volmer–Heyrovsky–Tafel mechanism and Volmer–Heyrovsky mechanism. *International Journal of Hydrogen Energy* 33 (9):2178–2188.

Boyjoo, Yash, Hongqi Sun, Jian Liu, Vishnu K. Pareek, and Shaobin Wang. 2017. A review on photocatalysis for air treatment: From catalyst development to reactor design. *Chemical Engineering Journal* 310:537–559.

Cai, Hairui, Bin Wang, Laifei Xiong, et al. 2020. Bridging effect of Co heteroatom between g-C_3N_4 and Pt NPs for enhanced photocatalytic hydrogen evolution. *Chemical Engineering Journal* 394:124964.

Cao, Shaowen, Baojia Shen, Tong Tong, Junwei Fu, and Jiaguo Yu. 2018. 2D/2D hetero-junction of ultrathin MXene/Bi_2WO_6 nanosheets for improved photocatalytic CO_2 reduction. *Advanced Functional Materials* 28 (21):1800136.

Cao, Shuang, Ting-Shan Chan, Ying-Rui Lu, et al. 2020. Photocatalytic pure water splitting with high efficiency and value by Pt/porous brookite TiO_2 nanoflutes. *Nano Energy* 67:104287.

Chen, Biao, Enzuo Liu, Fang He, et al. 2016. 2D sandwich-like carbon-coated ultrathin TiO_2@defect-rich MoS_2 hybrid nanosheets: Synergistic-effect-promoted electro-chemical performance for lithium ion batteries. *Nano Energy* 26:541–549.

Chen, Dongjie, Yanling Cheng, Nan Zhou, et al. 2020. Photocatalytic degradation of organic pollutants using TiO2-based photocatalysts: A review. Journal of Cleaner Production 268: 121725.

Chen, Lei, Yiming Xu, Zhi Yang, Kun Zhang, and Baoliang Chen. 2020. Cobalt (II)-based open-framework systems constructed on g-C_3N_4 for extraordinary enhancing photo-catalytic hydrogen evolution. Applied Catalysis B: Environmental 277: 119207.

Chen, Liuyun, Kelin Huang, Qingruo Xie, et al. 2021. The enhancement of photocatalytic CO_2 reduction by the in situ growth of TiO_2 on Ti_3C_2 MXene. *Catalysis Science & Technology* 11 (4):1602–1614.

Chen, Liuyun, Xinling Xie, Tongming Su, Hongbing Ji, and Zuzeng Qin. 2021. Co_3O_4/CdS p-n heterojunction for enhancing photocatalytic hydrogen production: Co-S bond as a bridge for electron transfer. *Applied Surface Science* 567:150849.

Chen, Sha, Danlian Huang, Piao Xu, et al. 2020. Semiconductor-based photocatalysts for photocatalytic and photoelectrochemical water splitting: Will we stop with photo-corrosion? *Journal of Materials Chemistry A* 8 (5):2286–2322.

Chen, Yilin, Xingchen He, Dongsheng Guo, et al. 2020. Supramolecular electrostatic self-assembly of mesoporous thin-walled graphitic carbon nitride microtubes for highly effi-cient visible-light photocatalytic activities. *Journal of Energy Chemistry* 49:214–223.

Cheng, Lei, Xin Li, Huaiwu Zhang, and Quanjun Xiang. 2019. Two-dimensional transition metal MXene-based photocatalysts for solar fuel generation. *The Journal of Physical Chemistry Letters* 10 (12):3488–3494.

Deng, Jiao, Pengju Ren, Dehui Deng, Liang Yu, Fan Yang, and Xinhe Bao. 2014. Highly active and durable non-precious-metal catalysts encapsulated in carbon nanotubes for hydrogen evolution reaction. *Energy & Environmental Science* 7 (6):1919–1923.

Ding, Li, Libo Li, Yanchang Liu, et al. 2020. Effective ion sieving with Ti_3C_2Tx MXene membranes for production of drinking water from seawater. *Nature Sustainability* 3 (4):296–302.

Dong, Hongjun, Shihuan Hong, Pingfan Zhang, et al. 2020. Metal-free Z-scheme 2D/2D VdW heterojunction for high-efficiency and durable photocatalytic H_2 production. *Chemical Engineering Journal* 395:125150.

Du, Chun, Qian Zhang, Zhaoyong Lin, Bo Yan, Congxin Xia, and Guowei Yang. 2019. Half-unit-cell $ZnIn_2S_4$ monolayer with sulfur vacancies for photocatalytic hydrogen evo-lution. *Applied Catalysis B: Environmental* 248:193–201.

Enyashin, A. N. and A. L. Ivanovskii. 2012. Atomic structure, comparative stability and electronic properties of hydroxylated Ti_2C and Ti_3C_2 nanotubes. *Computational and Theoretical Chemistry* 989:27–32.

Fang, Hongjun, Yusong Pan, Haixian Yan, et al. 2020. Facile preparation of Yb^{3+}/Tm^{3+} co-doped Ti_3C_2/Ag/Ag_3VO_4 composite with an efficient charge separation for boosting visible-light photocatalytic activity. *Applied Surface Science*:146909.

Fu, Fangbao, Dongjie Yang, Huan Wang, et al. 2019. Three-dimensional porous framework lignin-derived Carbon/ZnO composite fabricated by a facile electrostatic self-assembly showing good stability for high-performance supercapacitors. *ACS Sustainable Chemistry & Engineering* 7 (19):16419–16427.

Fujishima, Akira and Kenichi Honda. 1972. Electrochemical photolysis of water at a semiconductor electrode. *Nature* 238:37.

Ganesh, R. Sankar, M. Navaneethan, S. Ponnusamy, et al. 2018. Enhanced photon collection of high surface area carbonate-doped mesoporous TiO_2 nanospheres in dye sensitized solar cells. *Materials Research Bulletin* 101:353–362.

Ghidiu, M., M. R. Lukatskaya, M. Q. Zhao, Y. Gogotsi, and M. W. Barsoum. 2014. Conductive two-dimensional titanium carbide 'clay' with high volumetric capacitance. *Nature* 516 (7529):78–81.

Greiner, M. T., T. E. Jones, S. Beeg, et al. 2018. Free-atom-like d states in single-atom alloy catalysts. *Nature Chemistry* 10 (10):1008–1015.

Guo, Jing, Yingyuan Zhao, Anmin Liu, and Tingli Ma. 2019. Electrostatic self-assembly of 2D delaminated MXene (Ti_3C_2) onto Ni foam with superior electrochemical performance for supercapacitor. *Electrochimica Acta* 305:164–174.

Guo, Shaohui, Xuanhua Li, Jinmeng Zhu, Tengteng Tong, and Bingqing Wei. 2016. Au $NPs@MoS_2$ sub-micrometer sphere-ZnO nanorod hybrid structures for efficient photocatalytic hydrogen evolution with excellent stability. *Small* 12 (41):5692–5701.

Han, Xin, Lin An, Yue Hu, et al. 2020. Ti_3C_2 MXene-derived carbon-doped TiO_2 coupled with g-C_3N_4 as the visible-light photocatalysts for photocatalytic H_2 generation. *Applied Catalysis B: Environmental* 265:118539.

He, Binhong, Liang Chen, Mingjun Jing, Minjie Zhou, Zhaohui Hou, and Xiaobo Chen. 2018. 3D MoS_2-rGO@Mo nanohybrids for enhanced hydrogen evolution: The importance of the synergy on the Volmer reaction. *Electrochimica Acta* 283:357–365.

He, Nan, Xinwei Liu, Fei Gao, et al. 2020. Demonstration of 2D MXene memristor: Stability, conduction mechanism, and synaptic plasticity. *Materials Letters* 266:127413.

Hu, Peng, Chee Keong Ngaw, Yupeng Yuan, Prince Saurabh Bassi, Say Chye Joachim Loo, and Timothy Thatt Yang Tan. 2016. Bandgap engineering of ternary sulfide nanocrystals by solution proton alloying for efficient photocatalytic H_2 evolution. *Nano Energy* 26:577–585.

Huang, Guimei, Shuangzhi Li, Lijun Liu, Leifan Zhu, and Qiang Wang. 2020. Ti_3C_2 MXene-modified Bi_2WO_6 nanoplates for efficient photodegradation of volatile organic compounds. *Applied Surface Science* 503:144183.

Huang, Lixian, Bin Han, Xihe Huang, et al. 2019. Ultrathin 2D/2D $ZnIn_2S_4/MoS_2$ hybrids for boosted photocatalytic hydrogen evolution under visible light. *Journal of Alloys and Compounds* 798:553–559.

Irandost, Mozhgan, Rokhsareh Akbarzadeh, Meghdad Pirsaheb, Anvar Asadi, Parviz Mohammadi, and Mika Sillanpää. 2019. Fabrication of highly visible active N, S codoped $TiO_2@MoS_2$ heterojunction with synergistic effect for photocatalytic degradation of diclofenac: Mechanisms, modeling and degradation pathway. *Journal of Molecular Liquids* 291:111342.

Ji, Haiyan, Shan Shao, Guotao Yuan, et al. 2021. Unraveling the role of Ti_3C_2 MXene underlayer for enhanced photoelectrochemical water oxidation of hematite photoanodes. *Journal of Energy Chemistry* 52:147–154.

Jia, Guangri, Ying Wang, Xiaoqiang Cui, and Weitao Zheng. 2018. Highly carbon-doped TiO_2 derived from Mxene boosting the photocatalytic hydrogen evolution. *ACS Sustainable Chemistry & Engineering* 6 (10):13480–13486.

Jiang, Lisha, Jun Li, Kai Wang, Gaoke Zhang, Yuan Li, and Xiaoyong Wu. 2020. Low boiling point solvent mediated strategy to synthesize functionalized monolayer carbon nitride for superior photocatalytic hydrogen evolution. *Applied Catalysis B: Environmental* 260:118181.

Jiao, Yan, Yao Zheng, Mietek Jaroniec, and Shi Zhang Qiao. 2015. Design of electrocatalysts for oxygen- and hydrogen-involving energy conversion reactions. *Chemical Society Reviews* 44 (8):2060–2086.

Kong, Xianglong, Zhenbo Peng, Rui Jiang, et al. 2020. Nanolayered heterostructures of N-doped TiO_2 and N-doped carbon for hydrogen evolution. *ACS Applied Nano Materials* 3 (2):1373–1381.

Kuang, Panyong, Jingxiang Low, Bei Cheng, Jiaguo Yu, and Jiajie Fan. 2020. MXene-based photocatalysts. *Journal of Materials Science & Technology*.

Kulkarni, Aniruddha K., Rajendra P. Panmand, Yogesh A. Sethi, et al. 2018. In situ preparation of N doped orthorhombic Nb_2O_5 nanoplates /rGO composites for photocatalytic hydrogen generation under sunlight. *International Journal of Hydrogen Energy* 43 (43):19873–19884.

Kumaravel, Vignesh, Snehamol Mathew, John Bartlett, and Suresh C. Pillai. 2019. Photocatalytic hydrogen production using metal doped TiO_2: A review of recent advances. *Applied Catalysis B: Environmental* 244:1021–1064.

Li, Jinmao, Li Zhao, Shimin Wang, Jin Li, Guohong Wang, and Juan Wang. 2020. In situ fabrication of 2D/3D g-C_3N_4/Ti_3C_2 (MXene) heterojunction for efficient visible-light photocatalytic hydrogen evolution. *Applied Surface Science* 515:145922.

Li, Kunshan, Xinyu Lu, You Zhang, Kuiliang Liu, Yongchao Huang, and Hong Liu. 2020. Bi_3TaO_7/Ti_3C_2 heterojunctions for enhanced photocatalytic removal of water-borne contaminants. *Environmental Research* 185:109409.

Li, Weibing, Lin Wang, Qiang Zhang, et al. 2019. Fabrication of an ultrathin 2D/2D C_3N_4/MoS_2 heterojunction photocatalyst with enhanced photocatalytic performance. *Journal of Alloys and Compounds* 808:151681.

Li, Xiao, Yao Cheng, Qingping Wu, Ju Xu, and Yuansheng Wang. 2019. Synergistic effect of the rearranged sulfur vacancies and sulfur interstitials for 13-fold enhanced photocatalytic H_2 production over defective $Zn_2In_2S_5$ nanosheets. *Applied Catalysis B: Environmental* 240:270–276.

Li, Yang, Dainan Zhang, Xionghan Feng, Yulong Liao, Qiye Wen, and Quanjun Xiang. 2019. Truncated octahedral bipyramidal TiO_2/MXene Ti_3C_2 hybrids with enhanced photocatalytic H_2 production activity. *Nanoscale Advances* 1 (5):1812–1818.

Li, Yujie, Xiaotong Deng, Jian Tian, Zhangqian Liang, and Hongzhi Cui. 2018. Ti_3C_2 MXene-derived Ti_3C_2/TiO_2 nanoflowers for noble-metal-free photocatalytic overall water splitting. *Applied Materials Today* 13:217–227.

Li, Yujie, Lei Ding, Shujun Yin, et al. 2019. Photocatalytic H_2 evolution on TiO_2 assembled with Ti_3C_2 MXene and metallic 1T-WS_2 as co-catalysts. *Nano-Micro Letters* 12 (1):6.

Liao, Yuan, Jing Qian, Gang Xie, et al. 2020. 2D-layered Ti_3C_2 MXenes for promoted synthesis of NH_3 on P25 photocatalysts. *Applied Catalysis B: Environmental* 273:119054.

Ling, Chongyi, Li Shi, Yixin Ouyang, Qian Chen, and Jinlan Wang. 2016. Transition metal-promoted V_2CO_2 (MXenes): A new and highly active catalyst for hydrogen evolution reaction. *Advanced Science* 3 (11):1600180.

Liu, Jiapeng, Yizhe Liu, Danyun Xu, et al. 2019. Hierarchical "nanoroll" like MoS_2/$Ti_3C_2T_x$ hybrid with high electrocatalytic hydrogen evolution activity. *Applied Catalysis B: Environmental* 241:89–94.

Liu, Ning, Yun Chang, Yanlin Feng, et al. 2017. {101}–{001} Surface heterojunction-enhanced antibacterial activity of titanium dioxide nanocrystals under sunlight irradiation. *ACS Applied Materials & Interfaces* 9 (7):5907–5915.

Liu, Ning, Na Lu, HongTao Yu, Shuo Chen, and Xie Quan. 2020. Efficient day-night photocatalysis performance of 2D/2D Ti_3C_2/Porous g-C_3N_4 nanolayers composite and its application in the degradation of organic pollutants. *Chemosphere* 246:125760.

Liu, Peng, Wenjun Ding, Jing Liu, et al. 2020. Surface termination modification on high-conductivity MXene film for energy conversion. *Journal of Alloys and Compounds* 829:154634.

Liu, Yuanyue, Hai Xiao, and William A. Goddard. 2016. Schottky-barrier-free contacts with two-dimensional semiconductors by surface-engineered mxenes. *Journal of the American Chemical Society* 138 (49):15853–15856.

Liu, Yanyue, Jiaxin Yu, Dongfang Guo, Zijiong Li, and Yanjie Su. 2020. $Ti_3C_2T_x$ MXene/ graphene nanocomposites: Synthesis and application in electrochemical energy storage. *Journal of Alloys and Compounds* 815:152403.

Low, Jingxiang, Liuyang Zhang, Tong Tong, Baojia Shen, and Jiaguo Yu. 2018. TiO_2/ MXene Ti_3C_2 composite with excellent photocatalytic CO_2 reduction activity. *Journal of Catalysis* 361:255–266.

Ma, Qiansu, Xiaohong Hu, Na Liu, et al. 2020. Polyethylene glycol (PEG)-modified Ag/ Ag_2O/Ag_3PO_4/Bi_2WO_6 photocatalyst film with enhanced efficiency and stability under solar light. *Journal of Colloid and Interface Science* 569:101–113.

Manikandan, Arumugam, P. Robert Ilango, Chia-Wei Chen, et al. 2018. A superior dye adsorbent towards the hydrogen evolution reaction combining active sites and phase-engineering of (1T/2H) MoS_2/α-MoO_3 hybrid heterostructured nanoflowers. *Journal of Materials Chemistry A* 6 (31):15320–15329.

Meng, Aiyun, Jun Zhang, Difa Xu, Bei Cheng, and Jiaguo Yu. 2016. Enhanced photocatalytic H_2-production activity of anatase TiO_2 nanosheet by selectively depositing dual-cocatalysts on {101} and {001} facets. *Applied Catalysis B: Environmental* 198:286–294.

Naguib, Michael, Murat Kurtoglu, Volker Presser, et al. 2011. Two-dimensional nanocrystals produced by exfoliation of Ti_3AlC_2. *Advanced Materials* 23 (37):4248–4253.

Naguib, Michael, Vadym N. Mochalin, Michel W. Barsoum, and Yury Gogotsi. 2014. 25th anniversary article: MXenes: A new family of two-dimensional materials. *Advanced Materials* 26 (7):992–1005.

Ou, Wei, Jiaqi Pan, Yanyan Liu, et al. 2020. Two-dimensional ultrathin MoS_2-modified black Ti^{3+}–TiO_2 nanotubes for enhanced photocatalytic water splitting hydrogen production. *Journal of Energy Chemistry* 43:188–194.

Pandey, Mohnish, and Kristian S. Thygesen. 2017. Two-dimensional MXenes as catalysts for electrochemical hydrogen evolution: A computational screening study. *The Journal of Physical Chemistry C* 121 (25):13593–13598.

Peng, Chao, Ping Wei, Xiaoyao Li, et al. 2018. High efficiency photocatalytic hydrogen production over ternary Cu/TiO_2@$Ti_3C_2T_x$ enabled by low-work-function 2D titanium carbide. *Nano Energy* 53:97–107.

Peng, Chao, Xianfeng Yang, Yuhang Li, Hao Yu, Hongjuan Wang, and Feng Peng. 2016. Hybrids of two-dimensional Ti_3C_2 and TiO_2 exposing {001} facets toward enhanced photocatalytic activity. *ACS Applied Materials & Interfaces* 8 (9):6051–6060.

Qin, Yingying, Hong Li, Jian Lu, et al. 2020. Nitrogen-doped hydrogenated TiO_2 modified with CdS nanorods with enhanced optical absorption, charge separation and photocatalytic hydrogen evolution. *Chemical Engineering Journal* 384:123275.

Qin, Zuzeng, Liuyun Chen, Rujun Ma, et al. 2020. TiO_2/$BiYO_3$ composites for enhanced photocatalytic hydrogen production. *Journal of Alloys and Compounds* 836:155428.

Qu, Xiaofei, Meihua Liu, Zhaoqun Gao, et al. 2020. A novel ternary Bi_4NbO_8Cl/BiOCl/ Nb_2O_5 architecture via in-situ solvothermal-induced electron-trap with enhanced photocatalytic activities. *Applied Surface Science* 506:144688.

Ran, Jingrun, Guoping Gao, Fa-Tang Li, Tian-Yi Ma, Aijun Du, and Shi-Zhang Qiao. 2017. Ti_3C_2 MXene co-catalyst on metal sulfide photo-absorbers for enhanced visible-light photocatalytic hydrogen production. *Nature Communications* 8 (1):13907.

Rosenkranz, Andreas, Philipp G. Grützmacher, Rodrigo Espinoza, et al. 2019. Multi-layer $Ti_3C_2T_x$-nanoparticles (MXenes) as solid lubricants – Role of surface terminations and intercalated water. *Applied Surface Science* 494:13–21.

Shao, Jian, Weichen Sheng, Mingsong Wang, et al. 2017. In situ synthesis of carbon-doped TiO$_2$ single-crystal nanorods with a remarkably photocatalytic efficiency. *Applied Catalysis B: Environmental* 209:311–319.

Shi, Lang, Suqin Liu, and Zhen He. 2018. Construction of Sn/oxide g-C$_3$N$_4$ nanostructure by electrostatic self-assembly strategy with enhanced photocatalytic degradation performance. *Applied Surface Science* 457:1035–1043.

Shi, Xiaowei, Liang Mao, Ping Yang, et al. 2020. Ultrathin ZnIn$_2$S$_4$ nanosheets with active (110) facet exposure and efficient charge separation for cocatalyst free photocatalytic hydrogen evolution. *Applied Catalysis B: Environmental* 265:118616.

Si, Huayan, Qixin Deng, Chen Yin, et al. 2020. Gas exfoliation of graphitic carbon nitride to improve the photocatalytic hydrogen evolution of metal-free 2D/2D g-C$_3$N$_4$/graphdiyne heterojunction. *Journal of Alloys and Compounds* 833:155054.

Su, Tongming, Zachary D. Hood, Michael Naguib, et al. 2019a. 2D/2D heterojunction of Ti$_3$C$_2$/g-C$_3$N$_4$ nanosheets for enhanced photocatalytic hydrogen evolution. *Nanoscale* 11 (17):8138–8149.

Su, Tongming, Zachary D. Hood, Michael Naguib, et al.. 2019b. Monolayer Ti$_3$C$_2$T$_x$ as an effective co-catalyst for enhanced photocatalytic hydrogen production over TiO$_2$. *ACS Applied Energy Materials* 2 (7):4640–4651.

Su, Tongming, Rui Peng, Zachary D. Hood, et al. 2018. One-step synthesis of Nb$_2$O$_5$/C/Nb$_2$C (MXene) composites and their use as photocatalysts for hydrogen evolution. *ChemSusChem* 11 (4):688–699.

Su, Tongming, Qian Shao, Zuzeng Qin, Zhanhu Guo, and Zili Wu. 2018. Role of interfaces in two-dimensional photocatalyst for water splitting. *ACS Catalysis* 8 (3):2253–2276.

Sun, Jiwei, Shaorui Yang, Zhangqian Liang, et al. 2020. Two-dimensional/one-dimensional molybdenum sulfide (MoS$_2$) nanoflake/graphitic carbon nitride (g-C$_3$N$_4$) hollow nanotube photocatalyst for enhanced photocatalytic hydrogen production activity. *Journal of Colloid and Interface Science* 567:300–307.

Sun, Liming, Xiaoxiao He, Yusheng Yuan, et al. 2020. Tuning interfacial sequence between nitrogen-doped carbon layer and Au nanoparticles on metal-organic framework-derived TiO$_2$ to enhance photocatalytic hydrogen production. *Chemical Engineering Journal* 397:125468.

Sun, Liming, Yusheng Yuan, Fan Wang, Yanli Zhao, Wenwen Zhan, and Xiguang Han. 2020. Selective wet-chemical etching to create TiO$_2$@MOF frame heterostructure for efficient photocatalytic hydrogen evolution. *Nano Energy* 74:104909.

Sun, Shuchao, Jianjiao Zhang, Peng Gao, et al. 2017. Full visible-light absorption of TiO$_2$ nanotubes induced by anionic S$_2^{2-}$ doping and their greatly enhanced photocatalytic hydrogen production abilities. *Applied Catalysis B: Environmental* 206:168–174.

Sun, Yuliang, Di Jin, Yuan Sun, et al. 2018. g-C$_3$N$_4$/Ti$_3$C$_2$Tx (MXenes) composite with oxidized surface groups for efficient photocatalytic hydrogen evolution. *Journal of Materials Chemistry A* 6 (19):9124–9131.

Tang, Honghao, Huanran Feng, Huike Wang, Xiangjian Wan, Jiajie Liang, and Yongsheng Chen. 2019. Highly conducting MXene–silver nanowire transparent electrodes for flexible organic solar cells. *ACS Applied Materials & Interfaces* 11 (28):25330–25337.

Tang, Qijun, Zhuxing Sun, Shuang Deng, Haiqiang Wang, and Zhongbiao Wu. 2020. Decorating g-C$_3$N$_4$ with alkalinized Ti$_3$C$_2$ MXene for promoted photocatalytic CO$_2$ reduction performance. *Journal of Colloid and Interface Science* 564:406–417.

Tian, Pan, Xuan He, Lei Zhao, et al. 2019. Ti$_3$C$_2$ nanosheets modified Zr-MOFs with Schottky junction for boosting photocatalytic HER performance. *Solar Energy* 188:750–759.

Wang, Hou, Yuanmiao Sun, Yan Wu, et al. 2019. Electrical promotion of spatially photoinduced charge separation via interfacial-built-in quasi-alloying effect in hierarchical Zn$_2$In$_2$S$_5$/Ti$_3$C$_2$(O, OH)$_x$ hybrids toward efficient photocatalytic hydrogen evolution and environmental remediation. *Applied Catalysis B: Environmental* 245:290–301.

Wang, Jianhai, Yanfei Shen, Songqin Liu, and Yuanjian Zhang. 2020. Single 2D MXene precursor-derived TiO_2 nanosheets with a uniform decoration of amorphous carbon for enhancing photocatalytic water splitting. *Applied Catalysis B: Environmental* 270:118885.

Wang, Longyan, Yuanzhi Hong, Enli Liu, Xixin Duan, Xue Lin, and Junyou Shi. 2020. A bottom-up acidification strategy engineered ultrathin g-C_3N_4 nanosheets towards boosting photocatalytic hydrogen evolution. *Carbon* 163:234–243.

Wang, Peiyuan, Xiaoxuan Lu, Yash Boyjoo, et al. 2020. Pillar-free TiO_2/Ti_3C_2 composite with expanded interlayer spacing for high-capacity sodium ion batteries. *Journal of Power Sources* 451:227756.

Wang, Pengfei, Zhurui Shen, Yuguo Xia, et al. 2019. Atomic insights for optimum and excess doping in photocatalysis: A case study of few-layer Cu-$ZnIn_2S_4$. *Advanced Functional Materials* 29 (3):1807013.

Wang, Xinchen, Kazuhiko Maeda, Arne Thomas, et al. 2009. A metal-free polymeric photocatalyst for hydrogen production from water under visible light. *Nature Materials* 8 (1):76–80.

Wang, Yuxiong, Lei Rao, Peifang Wang, Zhenyu Shi, and Lixin Zhang. 2020. Photocatalytic activity of N-TiO_2/O-doped N vacancy g-C_3N_4 and the intermediates toxicity evaluation under tetracycline hydrochloride and Cr(VI) coexistence environment. *Applied Catalysis B: Environmental* 262:118308.

Wu, Zhibin, Yunshan Liang, Xingzhong Yuan, et al. 2020. MXene Ti_3C_2 derived Z–scheme photocatalyst of graphene layers anchored TiO_2/g–C_3N_4 for visible light photocatalytic degradation of refractory organic pollutants. *Chemical Engineering Journal* 394:124921.

Xiao, Rong, Chengxiao Zhao, Zhaoyong Zou, et al. 2020. In situ fabrication of 1D CdS nanorod/2D Ti_3C_2 MXene nanosheet Schottky heterojunction toward enhanced photocatalytic hydrogen evolution. *Applied Catalysis B: Environmental* 268:118382.

Xin, Xu, Yaru Song, Shaohui Guo, et al. 2020. In-situ growth of high-content 1T phase MoS_2 confined in the CuS nanoframe for efficient photocatalytic hydrogen evolution. *Applied Catalysis B: Environmental* 269:118773.

Xing, Zipeng, Jiaqi Zhang, Jiayi Cui, et al. 2018. Recent advances in floating TiO_2-based photocatalysts for environmental application. *Applied Catalysis B: Environmental* 225:452–467.

Xiong, Hailong, Lanlan Wu, Yu Liu, et al. 2019. Controllable synthesis of mesoporous TiO_2 polymorphs with tunable crystal structure for enhanced photocatalytic H_2 production. *Advanced Energy Materials* 9 (31):1901634.

Xu, Chong, Fan Yang, Bijian Deng, et al. 2020. $Ti3C_2/TiO_2$ nanowires with excellent photocatalytic performance for selective oxidation of aromatic alcohols to aldehydes. *Journal of Catalysis* 383:1–12.

Xu, Shiping, Xiang Sun, Yu Zhao, et al. 2018. Carbon-doped golden wattle-like TiO_2 microspheres with excellent visible light photocatalytic activity: Simultaneous in-situ carbon doping and single-crystal nanorod self-assembly. *Applied Surface Science* 448:78–87.

Xu, Yanjie, Shuai Wang, Jun Yang, et al. 2018. In-situ grown nanocrystal TiO_2 on 2D $Ti3C_2$ nanosheets for artificial photosynthesis of chemical fuels. *Nano Energy* 51:442–450.

Xu, Zhennan, Junxiang Jiang, Qianqian Zhang, Guobo Chen, Limin Zhou, and Ling Li. 2020. 3D graphene aerogel composite of 1D-2D Nb_2O_5-g-C_3N_4 heterojunction with excellent adsorption and visible-light photocatalytic performance. *Journal of Colloid and Interface Science* 563:131–138.

Yan, Junqing, Guangjun Wu, Naijia Guan, and Landong Li. 2014. Nb_2O_5/TiO_2 heterojunctions: Synthesis strategy and photocatalytic activity. *Applied Catalysis B: Environmental* 152-153:280–288.

Yang, Chengwu, Jiaqian Qin, Zhe Xue, Mingzhen Ma, Xinyu Zhang, and Riping Liu. 2017. Rational design of carbon-doped TiO_2 modified g-C_3N_4 via in-situ heat treatment for drastically improved photocatalytic hydrogen with excellent photostability. *Nano Energy* 41:1–9.

Yang, Chengwu, Xinyu Zhang, Jiaqian Qin, et al. 2017. Porous carbon-doped TiO_2 on TiC nanostructures for enhanced photocatalytic hydrogen production under visible light. *Journal of Catalysis* 347:36–44.

Yao, Xiaxi, Xiuli Hu, Wenjun Zhang, et al. 2020. Mie resonance in hollow nanoshells of ternary TiO_2-Au-CdS and enhanced photocatalytic hydrogen evolution. *Applied Catalysis B: Environmental* 276:119153.

Yi, Jianjian, Xiaojie She, Yanhua Song, et al. 2018. Solvothermal synthesis of metallic 1T-WS_2: A supporting co-catalyst on carbon nitride nanosheets toward photocatalytic hydrogen evolution. *Chemical Engineering Journal* 335:282–289.

Yin, Xing-Liang, Lei-Lei Li, Meng-Li Liu, Da-Cheng Li, Lei Shang, and Jian-Min Dou. 2019. MoS_x/CdS nano-heterostructures accurately constructed on the defects of CdS for efficient photocatalytic H_2 evolution under visible light irradiation. *Chemical Engineering Journal* 370:305–313.

Yuan, Nannan, Jinfeng Zhang, Sujuan Zhang, et al. 2020. What is the transfer mechanism of photoexcited charge carriers for g-C_3N_4/TiO_2 heterojunction photocatalysts? Verification of the relative p–n junction theory. *The Journal of Physical Chemistry C* 124 (16):8561–8575.

Yuan, Wenyu, Laifei Cheng, Yurong An, et al. 2018. Laminated hybrid junction of sulfur-doped TiO_2 and a carbon substrate derived from Ti_3C_2 MXenes: Toward highly visible light-driven photocatalytic hydrogen evolution. *Advanced Science* 5 (6):1700870.

Zhang, Chao, Yuming Zhou, Jiehua Bao, et al. 2018. Hierarchical honeycomb Br-, N-codoped TiO_2 with enhanced visible-light photocatalytic H_2 production. *ACS Applied Materials & Interfaces* 10 (22):18796–18804.

Zhang, Fan, Zirui Jia, Chao Wang, et al. 2020. Sandwich-like silicon/$Ti_3C_2T_x$ MXene composite by electrostatic self-assembly for high performance lithium ion battery. *Energy* 195:117047.

Zhao, Wei, Yue Feng, Haibao Huang, et al. 2019. A novel Z-scheme Ag_3VO_4/$BiVO_4$ heterojunction photocatalyst: Study on the excellent photocatalytic performance and photocatalytic mechanism. *Applied Catalysis B: Environmental* 245:448–458.

Zhong, Ruyi, Zisheng Zhang, Huqiang Yi, et al. 2018. Covalently bonded 2D/2D O-g-C_3N_4/TiO_2 heterojunction for enhanced visible-light photocatalytic hydrogen evolution. *Applied Catalysis B: Environmental* 237:1130–1138.

Zhou, Weijia, Zongyou Yin, Yaping Du, et al. 2013. Synthesis of few-layer MoS2 nanosheet-coated TiO_2 nanobelt heterostructures for enhanced photocatalytic activities. *Small* 9 (1):140–147.

Zhuang, Yan, Yunfei Liu, and Xianfeng Meng. 2019. Fabrication of TiO_2 nanofibers/MXene Ti_3C_2 nanocomposites for photocatalytic H_2 evolution by electrostatic self-assembly. *Applied Surface Science* 496:143647.

Zou, Yajun, Jian-Wen Shi, Dandan Ma, Zhaoyang Fan, Lu Lu, and Chunming Niu. 2017. In situ synthesis of C-doped TiO_2@g-C_3N_4 core-shell hollow nanospheres with enhanced visible-light photocatalytic activity for H_2 evolution. *Chemical Engineering Journal* 322:435–444.

Zou, Yajun, Jian-Wen Shi, Dandan Ma, Zhaoyang Fan, Chunming Niu, and Lianzhou Wang. 2017. Fabrication of g-C_3N_4/Au/C-TiO_2 hollow structures as visible-light-driven z-scheme photocatalysts with enhanced photocatalytic H_2 evolution. *ChemCatChem* 9 (19):3752–3761.

7 Application of MXene-Based Photocatalyst for Photocatalytic CO$_2$ Reduction

Ya Xiao and Tongming Su
School of Chemistry and Chemical Engineering, Guangxi University

CONTENTS

DOI: 10.1201/9781003156963-7

7.1   INTRODUCTION

With the remarkable growth of population, the demand for energy is increasing. It is estimated that the global energy demand is going to double by the middle of this century. Fossil fuels, including coal, oil, and natural gas, account for more than 80% of current energy demand (Nocera 2007). Such a heavy reliance on fossil fuels will not only accelerate the depletion of natural resources, but will also cause the release of large amounts of greenhouse gases, especially carbon dioxide (CO_2), thus disrupting the carbon balance in nature (Houghton 2005). Therefore, the establishment of effective methods to reduce the concentration of carbon dioxide in the atmosphere has become a worldwide research topic.

To this end, the facile and cost-effective capture and conversion of CO_2 into other chemical forms are of vital importance. During the past years, tremendous efforts have been done in this field. For example, carbon capture and storage (CCS) has been adopted to prevent the release of CO_2 gas into the atmosphere. However, high energy consumption and the risk of CO_2 leakage are two main disadvantages of this technology, thus limiting its widespread application. Catalytic hydrogenation of CO_2 to formate, formic acid, methanol, or dimethyl ether is another appealing approach for CO_2 conversion (Alvarez et al. 2017). However, current industrial-hydrogen (H_2) production methods, such as petroleum cracking, usually require a large amount of energy consumption. How to produce H_2 with a green and cost-effective pathway and cut down the energy input of driving this conversion currently are the most urgent issues to be solved.

In recent years, the insertion of CO_2 into organic substrates to generate high value-added carboxylic acid and carbonyl compounds is another hot topic of research (Johnson and Wendt 2014). However, the low conversion rate and the cost of raw materials limit its widespread application. Among various approaches, developing heterogeneous electrocatalysts such as metals, metal oxides, sulfides, and carbon-based materials to realize the valorization of CO_2 via electrocatalytic reduction appears to be an effective approach toward the production of value-added chemicals or fuels (Khezri, Fisher, and Pumera 2017). However, to realize efficient and selective CO_2 electroreduction is not easy because of the high overpotential of the cathodic process and the generation of multiple reaction intermediates (Zhang, Hu et al. 2018).

In nature, after absorbing solar energy by highly ordered pigments in photosynthetic organisms (such as green plants), highly energetic charge carriers will be generated (McConnell, Li, and Brudvig 2010). On one hand, the generated holes will be utilized in photosystem II (PSII) by oxygen evolving center (OEC) to realize water oxidation. On the other hand, the electrons will be transported to photosystem I (PSI) by an electron-transport chain and then mainly used in CO_2 fixation to produce carbohydrates, which has been recognized as the most important pathway to maintain the relative equilibrium of CO_2 and O_2 on the earth (Lima-Melo et al. 2019). Inspired by the natural model, one may wonder whether CO_2 can be captured and converted efficiently into useful chemicals or fuels through visible-light-induced catalysis under ambient temperature. With solar light as the mere energy input, the conversion of CO_2 into clean fuels (such as CO, CH_4, CH_3OH, CH_3CH_2OH, HCOOH, etc.) is an effective way to alleviate energy problems

(Han et al. 2018). The capture and conversion of CO$_2$ into chemicals and fuels can both mitigate the greenhouse effect and provide renewable energy. However, CO$_2$ is an inert and stable greenhouse gas due to the existence of covalent double bonds between carbon and oxygen atoms, which makes it difficult to activate and convert into value-added products (Park et al. 2022). The conversion of CO$_2$ requires high temperature, high pressure, or long-term reaction. Therefore, the development of functional materials with high CO$_2$ catalytic effect is a key factor for the efficient conversion of CO$_2$ under mild conditions. Photocatalytic reduction of CO$_2$ into valuable chemicals with a high-efficiency photocatalyst is a promising way to convert the solar energy to the chemical energy (Fu et al. 2019).

7.2 MECHANISM OF PHOTOCATALYTIC REDUCTION OF CO$_2$

From the view of thermodynamics, the reaction between CO$_2$ and water to produce organic matter and oxygen is a nonspontaneous process, and the reaction is feasible when a certain amount of energy is provided by the outside world (Hong et al. 2013). The process of photocatalytic reduction of CO$_2$ to synthesize cleaning products can be realized by using the photocatalytic materials under mild reaction conditions (normal temperature and pressure). In general, the typical photocatalytic reduction of CO$_2$ mainly includes three steps (White et al. 2015). Figure 7.1 shows the schematic of photocatalytic reduction CO$_2$: (1) Under the irradiation of light, photocatalysts absorb solar energy, and the incident photons with energy equal to or higher than the bandgap energy (Eg) of the photocatalyst can be absorbed to generate electron–hole (e^-–h^+) pairs. (2) The electrons and holes are separated from each other and spread to the surface of photocatalyst. In addition, they can also be transferred to the corresponding hole receptors and electron receptors to undergo redox reactions with the reactants. Due to the large number of defects and suspended bonds in the catalyst material, electrons or holes may be captured and redistributed. (3) The excited electrons have strong reductive property, which can cause CO$_2$ reduction reaction to produce hydrocarbons (such as CO, CH$_4$, CH$_3$OH, CH$_3$CH$_2$OH, HCOOH, etc.), while the holes at the valence band have strong oxidation property, which can cause the H$_2$O oxidation to produce O$_2$. Based on the total even numbers of H$^+$/e^- transferred in a complete reaction (from 2 to 8) to obtain different hydrocarbons, a deeper understanding of the reaction mechanism is essential for the design of innovative photocatalysts, especially when targeted to obtain specific products.

Usually, photocatalytic hydrogen-evolution reaction is ubiquitous in acidic solutions because there are more protons in the solution; CO$_2$ is easily converted to CO$_3^{2-}$ or HCO$_3^-$. Therefore, many studies for CO$_2$ reduction were conducted in the neutral electrolytes. Equations (7.1)–(7.9) show the potential to a normal hydrogen electrode (NHE) in an aqueous solution of pH = 7. According to different reaction mechanisms and pathways, CH$_4$, CO, CH$_3$OH, HCOOH, C$_2$H$_6$, C$_2$H$_5$OH, HCHO and CH$_3$COOH can be generated from the reaction of CO$_2$ reduction (Hong et al. 2013).

$$H_2O \rightarrow 0.5O_2 + 2H^+ + 2e^- \quad E = 0.82 \text{ V} \qquad (7.1)$$

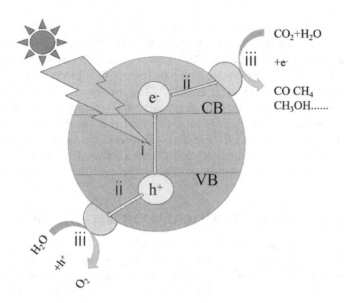

FIGURE 7.1 A schematic illustration of a probable mechanism for the photocatalytic reduction of CO_2. With the irradiation of light, incident photons with energy generate electron–hole (e^-–h^+) pairs. The electrons and holes are separated from each other and spread to the surface of the photocatalyst. Due to the large number of defects and suspended bonds in the catalyst material, electrons or holes may be captured and redistributed. The excited electrons cause CO_2 reduction reaction to produce hydrocarbons (such as CO, CH_4, CH_3OH, CH_3CH_2OH, HCOOH, etc.), while the holes at the valence band have strong oxidation property, which can cause the H_2O oxidation to produce O_2.

$$CO_2 + 8H^+ + 8e^- \rightarrow CH_4 + 2H_2O \quad E = -0.24 \text{ V} \tag{7.2}$$

$$CO_2 + 2H^+ + 2e^- \rightarrow CO + H_2O \quad E = -0.51 \text{ V} \tag{7.3}$$

$$CO_2 + 6H^+ + 6e^- \rightarrow CH_3OH + H_2O \quad E = -0.39 \text{ V} \tag{7.4}$$

$$2CO_2 + 12H^+ + 12e^- \rightarrow C_2H_5OH + 3H_2O \quad E = -0.33 \text{ V} \tag{7.5}$$

$$2CO_2 + 14H^+ + 14e^- \rightarrow C_2H_6 + 4H_2O \quad E = -0.27 \text{ V} \tag{7.6}$$

$$CO_2 + 4H^+ + 4e^- \rightarrow HCHO + H_2O \quad E = -0.55 \text{ V} \tag{7.7}$$

$$CO_2 + 2H^+ + 2e^- \rightarrow HCOOH \quad E = -0.58 \text{ V} \tag{7.8}$$

$$2CO_2 + 8H^+ + 8e^- \rightarrow CH_3COOH + 2H_2O \quad E = -0.31 \text{ V} \tag{7.9}$$

For an efficient photocatalyst, the following basic conditions need to be met. (1) The photocatalyst needs to have an appropriate position of the conduction band, valence band, and band gap. The reduction potential of conduction band of the photocatalyst must be more negative than the redox potential of CO_2 reduction (−0.24 eV), and the top of the valence band must be more positive than the redox potential of water oxidation (0.82 eV) (Wang and Zhang 2018). Due to the presence of overpotential, the band gap of a semiconductor photocatalyst must exceed the free energy of CO_2 (1.06 eV) (Teramura and Tanaka 2018). The electrons at the conduction band of the photocatalyst must have enough reducing ability to reduce CO_2, while holes at the valence band must have enough oxidation ability to oxidize H_2O. At the same time, a narrow band gap of the photocatalyst is needed to increase the absorption of visible light and improve the utilization of solar energy. (2) The photocatalyst is chemically stable and insoluble in water. (3) The preparation process is simple, and the photocatalysts are easy to obtain. (4) Photocatalytic activity is highly efficient and stable with good cycling performance. (5) It is easy to recycle, nontoxic, and harmless to the environment.

Since 1978, Halmann (Houghton 2005) reported a single crystal gallium phosphide, which can transform CO_2 into HCOOH, HCHO, and CH_4 under ultraviolet radiation. Semiconductor-based catalysts show the potential for photocatalytic CO_2 reduction. However, in order to improve the reaction efficiency and product selectivity, semiconductor catalytic materials need to be modified. (1) The energy level structure can be controlled to reduce the band gap so that more electrons can be generated and transferred to CO_2 molecules for reduction reaction under visible light. (2) Theoretically, it is important to control the size, active site, interface bond, and defect of photocatalyst to inhibit H_2 production and promote CO production at the same time. (3) They improve the separation efficiency of photogenerated charge carriers and extend the time of electrons and holes.

7.3 CO_2 CAPTURE BY MXENE-BASED MATERIALS

Transition metal carbides were used as potential materials for CO_2 capture (Giordano et al. 2009). The high CO_2 capture capacity of MXene could be ascribed to its high surface area, more adsorption sites, the rapid charge transfer from MXene to CO_2, and the high adsorption energy of CO_2 on the MXene surface (Morales-García et al. 2018). It was anticipated to achieve high CO_2 conversion yield by MXene because of the high CO_2 adsorption capacity, even at high temperatures and low CO_2 partial pressure. Improving the CO_2 adsorption on MXene can enhance the carbon dioxide activation and conversion by MXene-based materials.

7.3.1 SINGLE COMPONENT MXENE

The general formula of MXene is $M_{n+1}X_nT_x$ (M = Ti, Zr, Hf, V, Nb, Ta, Cr, Mo, W, n = 1, 2, 3, 4) (Anasori, Lukatskaya, and Gogotsi 2017). The CO_2 desorption was mainly affected by the adsorption strength and the adsorption conformation on the MXene. The highest CO_2 adsorption capacity could reach as high as 8.25 mole CO_2 per kilogram MXene. The CO_2 capture capacity of MXene is in the order of $Ti_2CTx > V_2CT_x > Zr_2CT_x > Nb_2CT_x > Mo_2CT_x > Hr_2CT_x > Ta_2CT_x > W_2CT_x$. The adsorption capacity of CO_2 on MXene was still high even at low CO_2 partial pressure and high temperature. Multiple sites would contribute to higher capacity of CO_2 capture, therefore increasing the ratio, size, and distribution of pores in MXene might be a favorable route to improve the CO_2 capture capacity.

Furthermore, increasing the specific area of MXene can enhance the CO_2 capture capacity. Specifically, the adsorption capacity of $Ti_3C_2T_x$ with a surface area of 21 $m^2 \cdot g^{-1}$ was 5.79 $mmol \cdot g^{-1}$, which was comparable to that of the common CO_2 absorbents. Particularly, the theoretical CO_2 capture capacity of MXene with the highest surface area can reach 44.2 $mmol \cdot g^{-1}$ (Wang et al. 2018). Therefore, designing and preparing the ultrathin MXene should be a feasible way to enhance the CO_2 adsorption capacity. Besides, the effect of atomic layers of MXene on the CO_2 capture was also investigated by using the density functional theory (Morales-Garcia et al. 2019).

The CO_2 adsorption capacity on the surface lone pair electrons of MXene was calculated by first principle calculation. MXenes, such as Ti_2CT_x, Zr_2CT_x, Hf_2CT_x, V_2CT_x, Nb_2CT_x, Ta_2CT_x, Cr_2CT_x, Mo_2CT_x, and W_2CT_x were selected for this study (Guo et al. 2020). Results showed that the Gibbs free energy of CO_2 adsorption on MXene was negative, which indicated that the CO_2 capture by MXene was thermodynamically favorable. Specifically, the value of Gibbs free energy of CO_2 adsorption by MXene was in the order of $Ti_2CT_x < Hf_2CT_x < Zr_2CT_x < V_2CT_x < Nb_2CT_x < Ta_2CT_x < Cr_2CT_x < Mo_2CT_x < W_2CT_x$. Interestingly, this order was relevant to the tendency of IVB > VB > VIB, which was correlated with the lone pair electron density of M atoms of MXene that transfers to CO_2.

7.3.2 ATOMIC DEFECTS OF MXENE

The ratio of atomic defect on MXene can influence the CO_2 capture capacity. Khaledialidusti et al. noticed that the intrinsic defect of $MO_2TiC_2T_x$ MXene favor the CO_2 adsorption (Khaledialidusti, Mishra, and Barnoush 2019). The synthesized structural defects were determined by the categories of terminators, as indicated by the theoretical formation energy of defect. For the defect formation, the defect generated on the outer layer (i.e., molybdenum layer) was easier than the inner layer (i.e., titanium layer). CO_2 was strongly absorbed on the defective sites of MXene via an exothermic process. Figure 7.2 discusses the reactivity of defected $Mo_2TiC_2O_2$ MXenes with a CO_2 molecule; Mo-containing MXenes are weakly reactive for CO_2 capture (Khaledialidusti, Mishra, and Barnoush 2020). Therefore, increasing the ratio of atomic defects on the MXene is considered as a efficient way to improve the CO_2 capture capacity.

In brief, the CO_2 conversion on the MXene-based materials can be enhanced by improving the CO_2 adsorption on MXene. Generally, CO_2 capture capacity can also be affected by external factors, such as temperature, pressure, and impurities. The structural factors for the improvement of CO_2 capture include the adsorption strength, adsorption conformation, number of interaction sites, and the ratio/size of pores in MXene-based hybrids. Absolutely, CO_2 capture capacity could also be enhanced by the functionalization of MXene and the introduction of CO_2−philic components in the MXene-based composite (Tang and Sun 2020).

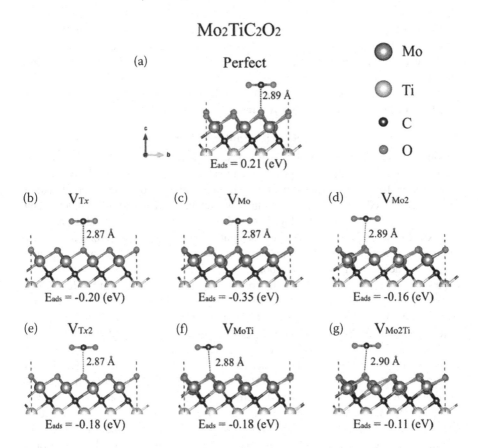

FIGURE 7.2 Relaxed adsorption configurations of a CO$_2$ molecule on the perfect and defected Mo$_2$TiC$_2$O$_2$ MXene. (a) Perfect MXene, (b) MXene–V$_{Tx}$, (c) MXene–V$_{Mo}$, (d) MXene–V$_{MO2}$, (e) MXene–V$_{Tx2}$, (f) MXene–V$_{MoTi}$, and (g) MXene–V$_{Mo2Ti}$ (Khaledialidusti, Mishra, and Barnoush 2020). The calculated results in the figure indicate that the CO$_2$ molecule is physisorbed on the perfect Mo$_2$TiC$_2$O$_2$ MXene, with a nonspontaneous reaction energy of 0.21 eV (see Fig. 5a). The carbon atom of CO$_2$ interacts with the oxygen surface functions and the molecule is placed at a distance of 2.89 Å while it is not significantly tilted. Moreover, a comparative nonspontaneous Gibbs free reaction energy (at 298.15 K) of 0.23 eV for the physisorption of a CO$_2$ molecule on the Mo$_3$C$_2$O$_2$ MXene was calculated.

7.4 CO$_2$ CONVERSION BY MXENE-BASED MATERIALS

7.4.1 MXene as Cocatalyst

In recent years, nitride metal, oxide metal salt, perovskite, and metal organic skeleton materials have been widely used as catalysts for CO$_2$ reduction (Peng et al. 2019). However, the photocatalytic reduction CO$_2$ efficiency is much lower than the requirement of practical application; this is mainly caused by the serious recombination of photogenerated electrons and holes. As a result, precious metal, such as Pt, is used to enhance the separation of the photogenerated electrons and holes (Khezri, Fisher, and Pumera 2017). However, the expensive cost is the main impediment for the practical use of noble metal. Using MXene as a co-catalyst is an effective means to improve the separation of the photogenerated charge carriers. When MXene is used as the co-catalyst, the photocatalytic performance of the photocatalyst can be significantly improved. Herein, we will discuss MXene as co-catalyst to promote the photocatalytic reduction of CO_2.

7.4.1.1 MXene/Nitride Composite Photocatalysts

Graphitic carbon nitride (g-C_3N_4) has aroused much concern in the photocatalytic reduction of CO_2 owing to its physicochemical stability, narrow band gap, appropriate energy band position, and the low cost. However, due to the fast recombination of photogenerated charge carriers and the low capability of adsorbing and activating CO_2 molecules, the g-C_3N_4 displayed poor photocatalytic performance (Jin-Shui, Bo, and Xin-Chen 2013). To enhance the separation efficiency of electrons and holes, multiple strategies were reported, including metal/nonmetal doping (Zhang, Zeai, and Li 2019), vacancy engineering (carbon, nitrogen vacancies (Tu et al. 2017)), molecular engineering (Yu et al. 2017), heterojunction construction (Xu et al. 2019), and so on. Recently, Ti_3C_2 as a co-catalyst was combined with g-C_3N_4 for photocatalytic hydrogen evolution, such as 2D/0D g-C_3N_4/Ti_3C_2 (He et al. 2020), 2D/2D g-C_3N_4/Ti_3C_2 (Dong et al. 2020), 2D/3D g-C_3N_4/Ti_3C_2 (Li, Zhao, et al. 2020). Due to the close contact between g-C_3N_4 and Ti_3C_2, the recombination of photogenerated carriers was effectively inhibited. Ti_3C_2 has a lower Fermi level (-0.53 eV vs. NHE) than the conduction band (-0.96 eV vs. NHE) of g-C_3N_4, which indicated that photogenerated electrons can freely transfer from g-C_3N_4 to Ti_3C_2 (Dong et al. 2020). Besides, the numerous hydrophilic functional groups (–OH, –O, –F) on the surface of Ti_3C_2 can establish a strong connection with g-C_3N_4; therefore, there is a compact contact interface between g-C_3N_4 and Ti_3C_2. In addition, the excellent electronic conductivity of Ti_3C_2 ensures effective charge-carrier transfer. These advantages make Ti_3C_2 a potential co-catalyst for heterojunction formation, and enhanced the photocatalytic performance of the photocatalyst.

2D/2D g-C_3N_4/Ti_3C_2 heterojunction has been reported for enhanced photocatalytic CO_2 reduction. Yang et al. (Yang et al. 2020) synthesize an ultrathin 2D/2D g-C_3N_4/Ti_3C_2 heterojunction by simple calcination of the mixture of multi-layered Ti_3C_2 particles and urea. For this method, urea not only acts as the gas template to exfoliate Ti_3C_2 into nanosheets, but it also acts as the precursor of g-C_3N_4 for in situ generating g-C_3N_4 nanosheets on the Ti_3C_2 surface. This method not only avoids the tedious process for the exfoliation of Ti_3C_2, but also benefits by the improved yield of ultrathin Ti_3C_2. It can be seen from Figure 7.3a that two kinds of nanosheets with the feature of transparence and smooth are stacked on top of each other. The partially enlarged region (blue box) clearly demonstrates the intimate interface contact between Ti_3C_2 and g-C_3N_4 nanosheets (Figure 7.3b). Under light irradiation, the excited electrons in the conduction band of g-C_3N_4 quickly migrate to the surface of Ti_3C_2. Subsequently, these accumulated electrons on the surface of Ti_3C_2 participated in the CO_2 reduction reaction. Meanwhile, H_2O was oxidized to O_2 by the holes in the VB of g-C_3N_4. The in situ FTIR results show the

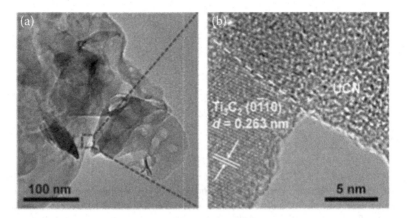

FIGURE 7.3 (a) TEM image of Ti$_3$C$_2$/g-C$_3$N$_4$ heterojunction (b) HRTEM image of the blue box region (Yang et al. 2020).

reaction process of photocatalytic CO$_2$ reduction. First, CO$_2$ molecules are adsorbed on the surface of Ti$_3$C$_2$/g-C$_3$N$_4$ samples, then they are transformed into HCOOH by the process of two-electron reduction, and finally, they generate CO (CO$_2$ + 2e$^-$ + 2 H$^+$ = CO + H$_2$O). Furthermore, HCOOH can be further reduced into HCHO and CH$_3$OH, finally into CH$_4$ (CO$_2$ + 8e$^-$ + 8H$^+$ = CH$_4$ + 2H$_2$O). The composite photocatalysts exhibit superior photocatalytic CO$_2$ reduction performance with the highest yield up to 5.19 μmol·h^{-1}·g^{-1} for CO and 0.044 μmol·h^{-1}·g^{-1} for CH$_4$. The total CO$_2$ conversion is improved about 8.1 times higher than that of g-C$_3$N$_4$. The significantly boosted photocatalytic performance originates from the formation of ultrathin 2D/2D Ti$_3$C$_2$/g-C$_3$N$_4$ heterojunction, which creates the intimate interface contact and quicker electron transfer, therefore promoting the spatial separation efficiency of photo-induced charge carriers.

Compared with the above Ti$_3$C$_2$, the alkalified Ti$_3$C$_2$ exhibited better photocatalytic reduction CO$_2$ property. Because the CO$_2$ molecule is an acidic oxide, an alkali sorbent is a benefit of the adsorption of CO$_2$ and the photocatalytic reduction CO$_2$ reaction. More –OH on the surface of alkalified Ti$_3$C$_2$ would result in better adsorption and activation of CO$_2$. For example, Tang et al. (Tang and Sun 2020) combined the g-C$_3$N$_4$ with alkalized Ti$_3$C$_2$ to study the effect of Ti$_3$C$_2$ as a noble metal-free co-catalyst on the photocatalytic CO$_2$ reduction. For this photocatalytic system, the CO evolution rate of the optimized sample (5% TCOH-CN) was 5.9 times higher than that of pure g-C$_3$N$_4$. The superior photocatalytic activity of 5% TCOH-CN was attributed to the excellent electrical conductivity and large CO$_2$ adsorption capacity of alkalized Ti$_3$C$_2$. The amounts of CO$_2$ adsorbed on Ti$_3$C$_2$ and alkalized Ti$_3$C$_2$ are 1.5 and 19.0 mg CO$_2$/g catalyst, respectively, at room temperature in 1 h, suggesting that KOH treatment can significantly increase the CO$_2$ adsorption on Ti$_3$C$_2$. The adsorption energy of CO$_2$ on Ti$_3$C$_2$ is −0.44 eV based on density functional theory (DFT) calculations, while that of CO$_2$ on alkalized Ti$_3$C$_2$ is −0.13 eV. That means the enhanced CO$_2$ adsorption amount on alkalized Ti$_3$C$_2$ can be ascribed to the lower adsorption energy of CO$_2$ on alkalized Ti$_3$C$_2$. In

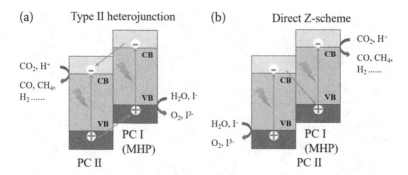

FIGURE 7.4 (a) Scheme showing the photocatalytic H_2 evolution and CO_2 reduction on dual semiconducting photocatalysts (PCs I and II) over Type-II (b) Direct Z-scheme photocatalyst. The direct Z-Scheme and the type II vertical structure are very similar; both are directed contact with two semiconductor materials, thus promoting photogenerated charge transfer quickly. In direct Z-scheme, the CB electrons of two kinds of semiconductor with lower energy combine with the VB hole, thus retaining the higher energy VB hole and CB electron, further maintaining the high Redox ability of the composite photocatalyst.

addition, 2D structure of the MXene tend to stack together by Van der Waals interaction, which will reduce the number of active centers. The preparation of the 3D MXenes structure has proven to be an effective strategy to reduce stack of the MXene layer, thus providing a larger specific surface area, higher porosity, and quality transmission distance.

The direct Z-scheme and type II system have become a hot research field topics because they do not need direct electron-transfer media (Raizada et al. 2019). Z-scheme and type II system (Figure 7.4) promote the photoinduced charge transfer in different ways due to the different charge-transfer driving-force level. The direct Z-scheme can maintain the high oxidation and reduction ability of the composite photocatalyst.

Tahir et al. (Tahir 2020) reported the pg-C_3N_4/Ti_3AlC_2/TiO_2 Z-Scheme heterogeneous junction system, which displayed a high efficiency for photocatalytic CO_2 reduction. The band-gap energy of the pg-C_3N_4/Ti_3AlC_2/TiO_2 Schottky catalyst was reduced to 2.98 eV as compared to TiO_2. The CB of TiO_2 (-0.52 eV) was more positive than that of the CB of pg-C_3N_4 (−1.17 eV). Through the Z-scheme heterojunction, electrons preferred to transfer from the CB of TiO_2 to the VB of pg-C_3N_4; due to the strong conductivity of Ti_3AlC_2, the electrons could be captured by Ti_3AlC_2 and then CO_2 reduction reaction occurs. In this system, methanol was not only the sacrificial reagent for generating electrons and trapping holes, but it also produced more protons (H^+), which would facilitate the photocatalytic reduction and produce hydrogen-rich products. In this way, the product selectivity was greatly improved.

The use of Ti_3C_2 improved the separation of the photogenerated carriers, as well as enhanced the stability of the hybrids. In addition, the dimensions and morphology of MXene can be adjusted to further enhance the photocatalytic CO_2 reduction.

7.4.1.2   MXene/Metal Oxide Composite Photocatalysts

As a typical semiconductor, TiO$_2$ is the most widely used photocatalyst for the photocatalytic reduction of CO$_2$ owing to its good chemical stability, environmental friendliness, and low cost (Meng et al. 2019a). However, it is still a challenge to achieve the practical application of the photocatalytic reduction of CO$_2$. This is mainly due to the low utilization of visible light, the fast recombination of photo-generated charge carriers, and the relatively wide band gap (3.2 eV) (Meng et al. 2019b). Over the recent years, numerous strategies have been investigated to improve the property of TiO$_2$, including increasing the specific surface area (Li, Wang et al. 2018), adding active sites for the adsorption of CO$_2$ (Liu et al. 2012), and so on. Numerous reports demonstrated that the Ti$_3$C$_2$ can offer abundant active sites when it was used as the co-catalyst for TiO$_2$, which led to the enhanced photocatalytic rate (Low et al. 2018).

Low et al. (Low et al. 2018) in situ produced TiO$_2$ nanoparticles on the Ti$_3$C$_2$ surface by a simple calcination method. The in situ growth method provided the intimate contact between the TiO$_2$ and the Ti$_3$C$_2$, and formed a unique rice crust-like structure, endowing it with large specific surface area. The loading amount of TiO$_2$ nanoparticles on Ti$_3$C$_2$ can be tailored by adjusting the calcination temperature. The high electronic conductivity of Ti$_3$C$_2$ is beneficial for the migration of photogenerated electrons from TiO$_2$ to Ti$_3$C$_2$, achieving spatial separation of photogenerated electron-hole pairs. As a result, the photocatalytic CO$_2$ reduction performance of the optimized TiO$_2$/Ti$_3$C$_2$ composite was 3.7 times higher than that of commercial TiO$_2$ (P25) due to the enlarged specific surface area and enhanced electron-hole separation efficiency. In conclusion, Ti$_3$C$_2$ as co-catalyst can extend the light absorption region and facilitate separation and transfer of photogenerated charge carriers on TiO$_2$. Hence, the photocatalytic reduction of CO$_2$ could be improved. However, it suffered from its own sets of limitations. For example, MXene is chemically unstable, and it can be easily oxidized in the humid air or water. According to the report, Ti$_3$C$_2$T$_x$ sheet was oxidized when it was stored in the water for seven days (Wan, Rajavel, Li 2020). In addition, the TiO$_2$ nanoparticles were found on the surface and the edge site of Ti$_3$C$_2$T$_x$. Therefore, the control of the transformation of TiO$_2$ to Ti$_3$C$_2$ is crucial to optimize the photocatalytic performance.

Conventional MXene usually exists in the form of 2D ultrathin nanosheet structure. When Ti$_3$C$_2$ turn from nanosheet to quantum dots, the Fermi level of Ti$_3$C$_2$ change from 0.71 V to −0.523 V. With the reduction of the Fermi level, photogenerated electrons can rapidly migrate from a semiconductor to Ti$_3$C$_2$ quantum dots, which realize the multi-electron CO$_2$ reduction reaction due to the enriched electrons on the Ti$_3$C$_2$. Ti$_3$C$_2$ quantum dots have different property compare to that of the two-dimensional materials. For example, Zeng et al. (Zeng et al. 2019) synthesized the Ti$_3$C$_2$ QDs/Cu$_2$O nanowires/Cu (NWs) heterostructure for the photocatalytic CO$_2$ reduction. The preparation process of this catalyst comprises two main steps. Firstly, the functionalized Ti$_3$C$_2$ QDs were obtained from Ti$_3$C$_2$ nanosheets by hydrothermal method and passivated by polyethylenimine (PEI). Secondly, the Ti$_3$C$_2$ QDs/Cu$_2$O nanowires/Cu (NWs) heterostructure was synthesized due to the electrostatic force between the positive charged PEI-Ti$_3$C$_2$ and negatively charged PSS-Cu$_2$O NWs/Cu.

The yield of CH_3OH over the Ti_3C_2 QDs/Cu_2O NWs/Cu is 8.25 and 2.15 times than those of Cu_2O NWs/Cu and Ti_3C_2 sheets/Cu_2O NWs/Cu after 6 h reaction.

In this Ti_3C_2 QDs/Cu_2O NWs/Cu system, the Ti_3C_2 QDs considerably enhanced the migration rate of the photoexcited electro-hole pairs in the host photocatalyst. The Ti_3C_2 QDs/Cu_2O nanowires/Cu (NWs) composite exhibited enhanced light absorption and smaller band gap compared to that of Ti_3C_2 sheets/Cu_2O NWs/Cu. Furthermore, Ti_3C_2 QDs prompt the migration of photogenerated electrons from Cu_2O NWs/Cu to Ti_3C_2 QDs. An energy-level structure diagram of the heterostructure system is illustrated in Figure 7.5. Under simulated solar light irradiation, electrons are excited from the VB of Cu_2O to its CB. As Ti_3C_2 QD is highly conductive and its E_F is less negative than the CB of Cu_2O, the photogenerated electrons tend to transfer to the Ti_3C_2 QD. Because the E_F of Ti_3C_2 QD is more negative than the redox potential of CO_2 to methanol, the accumulated electrons on Ti_3C_2 QD can stimulate this reduction reaction of CO_2. Moreover, Ti_3C_2 prefers to chemisorb CO_2 instead of H_2O, which further facilitates the CO_2 reduction. However, the reduction of CO_2 to CO cannot easily occur because the standard redox potential for this reaction (−0.53 V vs. NHE) is more negative than the Fermi level of Ti_3C_2 QDs. In conclusion, the enhanced CO_2 conversion efficiency and high selectivity can be assigned to the essential role of Ti_3C_2 QDs that promote light absorption, charge separation, transportation, and carrier density.

Cao et al. (Huang et al. 2020) synthesized a novel 2D/2D heterojunction of ultrathin Ti_3C_2/Bi_2WO_6 nanosheets by hydrothermal method, which showed significant improvements in the photocatalytic CO_2 reduction performance. The production rate of CH_4 and CH_3OH for Ti_3C_2/Bi_2WO_6 nanosheets was four times and six times higher than that of Bi_2WO_6 nanosheets. The enhancement of the photocatalytic performance can be attributed to the following advantages: (1) The photogenerated

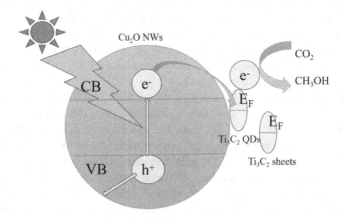

FIGURE 7.5 Energy level diagram of Ti_3C_2 QDs/Cu_2O NWs/Cu and Ti_3C_2 sheets/Cu_2O NWs/Cu heterostructures. Under simulated solar light irradiation, electrons are excited from the VB of Cu_2O to its CB. As Ti_3C_2 QD is highly conductive and its EF is less negative than the CB of Cu_2O, the photogenerated electrons tend to transfer to the Ti_3C_2 QD. The EF of Ti_3C_2 QD is more negative than the redox potential of CO_2 to methanol, the accumulated electrons on Ti_3C_2 QD can stimulate this reduction reaction of CO_2.

electrons on the CB of Bi$_2$WO$_6$ could rapidly transfer to the surface of Ti$_3$C$_2$ through the 2D/2D compact interface. The large contact interface could shorten the transport distance of the charge carriers. (2) The existence of Ti$_3$C$_2$ increases the CO$_2$ adsorption capability. (3) The photothermal effect of Ti$_3$C$_2$ is able to activate the catalyst and impel the photocatalytic reaction. Moreover, the photogenerated electrons can quickly transfer from Bi$_2$WO$_6$ to the Ti$_3$C$_2$ via the 2D/2D heterojunction interface. Therefore, the electrons will accumulate on the surface of Ti$_3$C$_2$, and the photocatalytic reduction efficiency of CO$_2$ can be effectively enhanced.

7.4.1.3 MXene/Perovskite Composite Photocatalysts

Recently, all-inorganic cesium lead halide (CsPbX$_3$, X = Cl, Br, I) perovskite nanocrystals (NCs) have received enormous attention due to their unique optoelectronic properties, such as high photoluminescence efficiencies (> 90%), a wide range of colors, narrow emission bands, and highly tunable band gap. However, their lack of active sites limits their redox capabilities when used as photocatalysts. Pan et al. (Intikhab et al. 2019) prepare CsPbBr$_3$/Ti$_3$C$_2$ nanocomposites and used for the photocatalytic CO$_2$ reduction reaction. CsPbBr$_3$/Ti$_3$C$_2$ nanocomposites were prepared by in situ growing the CsPbBr$_3$ nanocrystals on the Ti$_3$C$_2$ nanosheets. CsPbBr$_3$/MXene-20 nanocomposites exhibited the highest CO and CH$_4$ generation rates of 26.32 and 7.25 μmol·g^{-1}·h^{-1}, respectively, which were higher than that of CsPbBr$_3$ NCs (< 4.4 μmol·g^{-1}·h^{-1}). The CB offset (1.5 eV) between them could effectively promote the transfer the photogenerated electrons to MXene nanosheets and enhanced the separation of photogenerated electrons in CsPbBr$_3$ NCs. Moreover, the Schottky barrier between CsPbBr$_3$ NCs and MXene nanosheets prevented the flow of photogenerated electrons back to the surface of the CsPbBr$_3$ NCs (Figure 7.6). In short, MXene can enhance the separation of the photogenerated electrons and holes on the CsPbBr$_3$ NCs, thus improving the photocatalytic CO$_2$ reduction.

7.4.1.4 MXene/Metal-Organic Framework Composite Photocatalysts

Metal-organic frameworks (MOF) is a new type of porous material composed of metal ions and organic ligands. With the advantages of high porosity, low density,

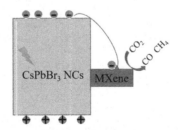

FIGURE 7.6 Schematic illustration of the relative energetic diagram of the CsPbBr$_3$ NCs/MXene-n heterostructures suitable for photocatalytic CO$_2$ reduction. The Schottky barrier between CsPbBr$_3$ NCs and MXene nanosheets prevented the flow of photogenerated electrons back to the surface of the CsPbBr$_3$ NCs. MXene can enhance the separation of the photogenerated electrons and holes on the CsPbBr$_3$ NCs. The excited electrons cause CO$_2$ reduction reaction to produce hydrocarbons (such as CO, CH$_4$).

large specific surface area, regular pore channels, adjustable pore size, and topological structure diversity, MOF have been widely used as photocatalysts for the photocatalytic CO_2 reduction (Wang et al. 2020). The photocatalytic activity of MOFs can be enhanced by coupling with the Ti_3C_2, and the enhancement can be ascribed to the formation of the Schottky junction between MOF and Ti_3C_2 (Zhao et al. 2019). Due to the Schottky junction, the recombination of photogenerated electrons and holes was suppressed, and the photogenerated electrons could accumulate on the surface of the Ti_3C_2 nanosheet for the photocatalytic CO_2 reduction.

Layered double hydroxides (LDH) have also attracted great attention in the field of photocatalytic CO_2 reduction due to their alkalinity and strong adsorption capacity (Chen et al. 2020). For example, Chen et al. (Chen et al. 2020) reported Co-Co LDH/MXene nanoarrays via an in situ MOF-derived strategy. In the synthetic process, ZIF-67 MOFs turn into vertical 2D Co-Co LDH on Ti_3C_2 nanosheets and form the Co-Co LDH/Ti_3C_2 nanosheet composite (Co-Co LDH/TNS). For the Co-Co LDH/TNS photocatalyst, the photocatalytic CO_2 reduction performance can be improved by exposing more catalytically active sites and the acceleration of the electron transport to the active sites. The CO production rate of the Co-Co LDH/TNS-15 sample was 2.2 times higher than that of the Co-Co LDH.

The photocatalytic CO_2 reduction mechanism of the Co-Co LDH/TNS composite with [Ru(bpy)$_3$]Cl$_2$ as the photosensitizer is shown in Figure 7.7. First, the Ru-based photosensitizer was first transformed into the excited state ([Ru(bpy)$_3$]Cl$_2$*)

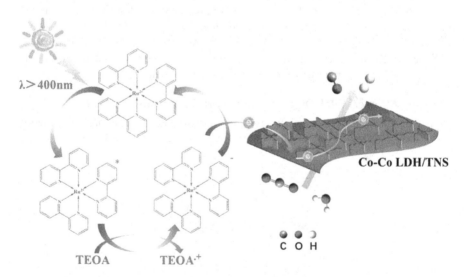

FIGURE 7.7 Proposed photocatalytic mechanism of the Co–Co LDH/TNS composite with [Ru(bpy)$_3$]Cl$_2$ for the visible-light-driven photocatalytic reduction of CO_2 (Chen et al. 2020). Firstly, Ru-based photosensitizer was firstly transformed into the excited state ([Ru (bpy)$_3$]Cl$_2$*) under the visible-light irradiation, followed by the quench via sacrificial electron donor (TEOA) to generate reduced photosensitizer. Secondly, the electrons generated on the [Ru(bpy)$_3$]Cl$_2$– transfer to the Co-Co LDH/TNS hybrids. Finally, the active site (i.e., Co) received the transferred electrons for the CO_2 reduction. Meanwhile, the proton in the system was also reduced to H_2 by the photogenerated electrons.

under the visible-light irradiation, followed by the quench via sacrificial electron donor (TEOA) to generate reduced photosensitizer. Second, the electrons generated on the [Ru(bpy)$_3$]Cl$_2^-$ transfer to the Co-Co LDH/TNS hybrids. Finally, the active site (i.e., Co) received the transferred electrons for the CO$_2$ reduction. Meanwhile, the proton in the system was also reduced to H$_2$ by the photogenerated electrons. For this photocatalytic system, the flat band potential of Co-Co@MXene composite (–1.01 V vs. NHE) was lower than that of redox potential value of Ru-based photosensitizer (–1.09 V vs. NHE), indicating that the electron could transfer from the latter to the former spontaneously for the reduction of CO$_2$ to CO. Moreover, the conduction band minimum (i.e., 3.33 eV) of Co-Co@MXene composite was between the LUMO (3.19 eV) and HOMO (5.68 eV) energy of Ru-based photosensitizer. Therefore, the photoinduced electrons in the Ru-based photosensitizer could be favorably transmitted into conduction band of Co-Co@MXene composite, and triggered the CO$_2$ reduction reaction.

Table 7.1 shows the photocatalytic CO$_2$ reduction performance of several photocatalysts before and after combining with MXene. Notably, the photocatalytic performance of photocatalysts was significantly improved after using MXene as the co-catalyst. In addition, the main product of the photocatalytic CO$_2$ reduction is CO and CH$_4$, and with CH$_3$OH, C$_2$H$_4$, CH$_3$CHO as the by-products. However, C2 product is difficult to obtain in high efficiency and high selectivity on the MXene-based composites. At present, there are few studies on deep understanding the intrinsic catalytic mechanism of MXenes. Previous works reveal that the catalytic performance of MXenes is highly dependent on its surface defects (Khaledialidusti, Mishra, and Barnoush 2019), termination groups (Seh et al. 2016), and electronic configurations (Intikhab et al. 2019), which motivates great research efforts dedicated to tailor the structural properties of MXenes for improving their catalytic performance.

7.4.2 THE EFFECT OF SURFACE FUNCTIONAL GROUPS OF MXENE ON THE PHOTOCATALYTIC CO$_2$ REDUCTION

MXenes are prepared by etching the MAX phase with various etchant, such as HF and LiF/HCl. Therefore, different functional groups were generated on the surface of MXene (–OH, –F, –O, etc.) (Ghidiu et al. 2014). The surface chemical property of MXenes is closely related to the surface-functional groups. For example, surface-terminal groups can provide a large number of active sites for the CO$_2$ adsorption and reaction, which inspired researchers to modify the surface-termination groups of MXenes to obtain the highly active catalyst. So far, the synthesis of MXenes with uniform and pure surface-termination groups is still an experimental challenge. MXene with mixed functional groups has attracted great attention.

The distribution of the surface-functional groups largely depends on the synthetic conditions, such as the etching agent, pH, and temperature. For example, the ratio of O/OH on the surface of MXene depends on the pH (Natu, Sokol, and Verger 2018). At low pH the favored termination is OH, increasing the pH leads to dihydroxylation of the OH into O. In addition, different steps of water centrifugation increase the pH to 5, resulting in a partial dihydroxylation and the formation of O terminal groups on

TABLE 7.1

Summary of Photocatalyst for the Photocatalytic CO_2 Reduction before and after Combining with MXene

Photocatalyst	Dosage	Light Source	Reaction Conditions	Yields of Products	Ref.
2D g-C_3N_4	20 mg	300 W Xe lamp	NaHCO$_3$, H$_2$SO$_4$	CO, 0.62 µmol·g^{-1}·h^{-1} CH$_4$, 0.021 µmol·g^{-1}·h^{-1}	(Yang et al. 2020)
2D-2D g-C_3N_4/Ti$_3$C$_2$			CO$_2$ and H$_2$O vapor	CO, 5.19 µmol·g^{-1}·h^{-1} CH$_4$, 0.044 µmol·g^{-1}·h^{-1}	(Tang and Sun 2020)
g-C_3N_4	40 mg	300 W Xe lamp	H$_2$O vapor	CO, 1.88 µmol·g^{-1}·h^{-1} CH$_4$, 0.048 µmol·g^{-1}	
g-C_3N_4/alkalized Ti$_3$C$_2$				CO, 11.21 µmol·g^{-1} CH$_4$, 0.269 µmol·g^{-1}	
TiO$_2$	50 mg	300 W Xe lamp	NaHCO$_3$, HCl	CH$_4$, 0.06 µmol·h^{-1}	(Low et al. 2018)
TiO$_2$/Ti$_3$C$_2$			CO$_2$, H$_2$O vapor	CH$_4$, 0.22 µmol·h^{-1}	
Cu$_2$O NWs/Cu	2 cm^2	300 W Xe lamp	H$_2$O vapor	CH$_3$OH, 2.95 ppm·cm^{-2}·h^{-1}	(Zeng et al. 2019)
Cu$_2$O/Ti$_3$C$_2$ sheets/Cu				CH$_3$OH, 11.98 ppm·cm^{-2}·h^{-1}	
Cu$_2$O/Ti$_3$C$_2$ QDs/Cu				CH$_3$OH, 25.77 ppm·cm^{-2}·h^{-1}	
Bi$_2$WO$_6$	100 mg	300 W Xe lamp	NaHCO$_3$, H$_2$SO$_4$	CH$_4$, 0.41 µmol·g^{-1}·h^{-1} CH$_3$OH, 0.07 µmol·g^{-1}·h^{-1}	(Huang et al. 2020)
Bi$_2$WO$_6$/Ti$_3$C$_2$			CO$_2$, H$_2$O vapor	CH$_4$, 1.78 µmol·g^{-1}·h^{-1} CH$_3$OH, 0.44 µmol·g^{-1}·h^{-1}	
CsPbBr$_3$	–	300 W Xe lamp	Ethyl acetate	CO, < 4.4 µmol·g^{-1}·h^{-1} CH$_4$, < 4.4 µmol·g^{-1}·h^{-1}	(Intikhab et al. 2019)
CsPbBr$_3$/Ti$_3$C$_2$				CO, 26.32 µmol·g^{-1}·h^{-1} CH$_4$, 7.25 µmol·g^{-1}·h^{-1}	
Co-Co LDH	0.5 mg	5 W LED	MeCN/H$_2$O/triethanolamine (TEOA), Ru(bpy)$_3$]Cl$_2$·6H$_2$O	CO, 2.84 µmol·h^{-1}	(Chen et al. 2020)
Co-Co LDH/Ti$_3$C$_2$				CO, 6.248 µmol·h^{-1}	

the MXene surfaces. In particular, the amount of F terminal on MXenes obtained with HF etching agent is nearly four times higher than LiF/HCl etching agent.

Ingemar et al. (Persson et al. 2017) investigated the effect of temperature on the surface groups on $Ti_3C_2T_x$ MXene. The temperature-programmed XPS (TP-XPS) showed that the functional groups consist of F and O occupying well-defined sites on the Ti_3C_2 surfaces. While the functional F coordinates to a single site, the functional O coordinates to two sites. As the temperature exceeds 550 °C and increase to 750 °C, the functional F was gradually desorbed and removed. With the desorption of F, the O surface atoms coordinates to the alternative site, which most likely is a bridge position, migrate to the preferred coordination vacated by the desorbed F. In situ heating of a few-layered $Ti_3C_2T_x$ sheets, the initial disorder surface transforms into highly ordered structures at high temperatures, which corresponding to the reconfiguration temperatures observed from the TP-XPS. In addition, the functional O atoms were found to thermodynamically favor the A-sites, on top of the Ti atoms of the middle layer, which was confirmed by the high-resolution STEM imaging and corroborating STEM image simulations. Hence, in the heating process, part of MXene surface was bare and unfunctionalized; therefore, further functionalization can be introduced. The heating process can be a way to manipulate the functionalization of MXene surface.

7.4.2.1 Tuning the Adsorption Energy of Reaction Intermediates

MXene with different functional groups has been synthesized and widely used for photocatalytic reaction. Aleksandra et al. (Yoon et al. 2018) studied the impact of –F termination on the photocatalytic hydrogen evolution (HER) performance of MXenes. They use KOH to synthesize low –F termination by replace –F with –O or –OH functional groups. The density functional theory (DFT) calculation confirmed that the high concentration of –F termination could suppress the HER activity of $Ti_3C_2T_x$. Su et al. (Jiang, Tao, and Xi 2019) further transformed –OH into –O on $Ti_3C_2T_x$ via the alkaline treatment and calcination at 450 °C under the inert gas environment. The O-terminated Ti_3C_2 nanosheets show a better HER kinetics activity. These results strongly suggest that engineering the termination groups of MXenes can directly tune the activity of $Ti_3C_2T_x$.

The N doping and Ti vacancy (V_{Ti}) on Ti_3C_2 MXene can stabilize the key *COOH intermediates and accelerates the desorption of *CO, resulting in excellent CO_2 reduction activity. Qu et al. (Qu et al. 2020) synthesized N-doped Ti_3C_2 MXene nanosheets with abundant V_{Ti} defects (NTC-V_{Ti}) by a facile NH_3-assisted pyrolysis method. For the photocatalytic CO_2 reduction reaction, the hydrogen evolution reaction (HER) usually competes with the photocatalytic CO_2 reduction reaction (CO_2RR), which leads to the poor selectivity for the CO_2RR. Based on the DFT calculation, NTC-V_{Ti} exhibits a smaller negative free-energy change for *H adsorption (−0.19 eV) than that of TC (−0.13 eV), confirming that the HER on NTC-V_{Ti} is effectively inhibited (Figure 7.8a). The free energy profiles of the CO_2RR catalyzed by NTC-V_{Ti} and TC are presented in Figure 7.8b. It can be seen that NTC-V_{Ti} shows a free-energy change of −0.01 eV, lower than that of TC (0.02 eV), suggesting that CO_2 adsorption is easier on NTC-V_{Ti}. Compared with TC, there is an improvement in the free-energy change for the first hydrogenation step (0.14 eV) and a decrease of CO desorption free-energy change (0.11 eV) for

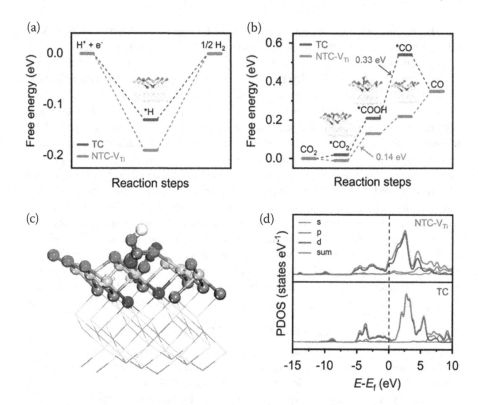

FIGURE 7.8 Mechanistic calculations: (a) Calculated free energy of hydrogen adsorption. (b) Calculated free energy of the CO_2RR; inset: the corresponding optimized adsorption configurations of intermediates on NTC-V_{Ti} . (c) Electron density difference for *COOH adsorbed on NTC-V_{Ti}. Note that the purple and yellow colors represent the electron accumulation and electron depletion, respectively. (d) Calculated PDOS of NTC-V_{Ti} and TC, corrected with the Fermi level. (Qu et al. 2020) Copyright (2020) Nanoscale. DFT calculation: NTC-VTi exhibits a smaller negative free-energy change for *H adsorption (−0.19 eV) than that of TC (−0.13 eV) (Figure 7.8a). The free energy profiles of the CO_2RR catalyzed by NTC-VTi and TC are presented in Figure 7.8b. NTC-VTi has a free-energy change of −0.01eV, lower than that of TC (0.02 eV), suggesting that CO_2 adsorption is easier on NTC-VTi. Compared with TC, there is an improvement in the free-energy change for the first hydrogenation step (0.14 eV) and a decrease of CO desorption free-energy change (0.11 eV) for NTC-VTi. As a result, the selective CO_2-to-CO pathway on NTC-VTi is much more effective than that on TC. The DFT-based electron density of NTC-VTi with *COOH adsorption is depicted in Figure 7.8c, which shows that the adsorption state of *COOH can hybridize with the surface state of Ti sites in NTC-VTi, and thus stabilize the *COOH intermediate. NTC-VTi displayed higher occupied states than TC near the Fermi level, which can enhance the electron transfer from the active center to the reactants and intermediates (Figure 7.8d).

NTC-V_{Ti}. As a result, the selective CO_2-to-CO pathway on NTC-V_{Ti} is much more effective than that on TC. The DFT-based electron density of NTC-V_{Ti} with *COOH adsorption is depicted in Figure 7.8c, which shows that the adsorption state of *COOH can hybridize with the surface state of Ti sites in NTC-V_{Ti}, and thus stabilize the *COOH intermediate. Moreover, as revealed by the projected density of states in Figure 7.8d, NTC-V$_{Ti}$ displayed higher occupied states than TC near the Fermi level, which can enhance the electron transfer from the active center to the reactants and intermediates. Therefore, stabilizing the key *COOH intermediate and enhancing the conductivity induced by N dopants and V$_{Ti}$ defects can synergistically optimize the reaction route and boost the CO$_2$RR performance.

7.4.2.2 Tuning the Electronic State of Adjacent Atoms

The terminated functional groups can also affect the electronic structure of adjacent metal atoms on MXenes, thus affecting the catalytic performance. Although the terminations on MXene surface did not directly show the catalytic activity, it was found that the −F termination can weaken adsorption of atomic hydrogen on its adjacent Pd sites in Pd/Nb$_2$C, leading to a favorable ΔG_{H*} for HER (Zhang et al. 2020). Therefore, different termination groups can be rationally introduced to tune the catalytic performance of their adjacent metal sites on MXenes.

7.4.3 ATOMIC ENGINEERING

Atomic engineering involves the synthesis of the single atom catalyst or doping of heteroatoms into the crystal structures to achieve desirable properties. Engineering MXene at the atomic level has been widely studied to adjust its catalytic property, which can be divided into two categories according to the catalytic active sites. One is to adjust the lattice geometry by inserting the extra atoms into the crystal structure of MXene (Yoon et al. 2018); the other one is to introduce the extra highly active atoms on the MXene (Qu et al. 2018).

7.4.3.1 Adjusting the Lattice Structure

The fundamental property of MXene is closely related to its lattice structure, which can be adjusted to achieve high catalytic property. For example, Ti$_3$C$_{1.6}$N$_{0.4}$ MXene derived from the Ti$_3$AlC$_x$N$_{2-x}$ MAX exhibited higher electrical conductivity (81.74 ± 1.16 mS·cm^{-1}) than that of Ti$_3$C$_{1.8}$N$_{0.2}$ (28.42 ± 0.43 mS·cm^{-1}) and Ti$_3$C$_2$ (21.90 ± 0.39 mS·cm^{-1}) (Tang et al. 2020). Therefore, the electrical conductivity of MXene can be enhanced by adjusting the lattice structure, which can accelerate transfer of photogenerated electrons and enhance the performance of the photocatalytic reaction.

7.4.3.2 Introducing Extra Active Center

Vacancy can improve the performance of catalysts in two aspects. First, the electron density around the vacancy benefits the affinity of CO$_2$ molecules. Second, some vacancies can be used as electron-capture sites to promote the separation of charge carriers.

Monolayer or multilayer MXenes was usually prepared by etching the Al layer of MAX with HF solution. In this way, even in the mild condition, the some surface Ti atoms can be etched away, leading to the formation of the single Ti vacancy or vacancy clusters on the MXene surface (Alhabeb et al. 2017). Interestingly, the Ti vacancy on MXene is very unstable and active, which makes the MXene highly reductive. Therefore, when metal salt precursor is contact with the MXene, the metal salt can be reduced on the MXene surface without any reductant (Zhao and Yao 2020). In addition, Ti defects on MXene was considered as an ideal position to stabilize the single metal atoms. After the removal of Al layer of MAX, surface functional groups, such as –O, –OH, –F is beneficial to the electrostatic adsorption of metal precursor, which favors the subsequent reduction and anchoring of metal atoms. When single atoms are reduced on the position of Ti defect by in situ self-reduction process on the vacancy-rich MXene at room temperature, strong metal-carbon bonds can be formed on the MXene, which is conducive to the stability of the metal atom. To sum up, MXene with suitable surface defect and high reductive ability is expected to be an ideal carrier for the preparing the single atom catalyst.

Many studies have shown that reaction sites on the substrate can be doped with single atom to maximize the catalytic efficiency. When the individual atoms are doped into the lattice, the physical/chemical property of the catalyst will be changed due to the different valence, ion radius, and electron density of the doped atom. In general, when extra atoms expose on the photocatalyst surface, they can act as the reactive sites and affect the adsorption of the reactants (Gao et al. 2016). The different adsorption energy of reactants can affect the way of redox reaction and produce different products. For example, the adsorption of CO_2 molecules can be enhanced by doping with the metal single atom, and the doped single atom can form the reaction sites for photocatalytic CO_2 reduction (Yuan et al. 2019). In addition, when single atom dispersed in the crystal lattice, these single atoms unsaturated coordination with the adjacent atom, which can accumulate the electrons and provide enough electrons for CO_2 reduction.

The aggregation of electrons at reactive sites plays an important role on the photocatalytic performance. Reaction sites induced by the doped atoms can accept photogenerated electrons from the matrix material or photosensitizer, then these electrons can transfer to the reactants, such as CO_2, H_2O, and intermediates, for the redox reaction. Photocatalytic reduction of CO_2 is complicated, the production of CH_4 is a process of eight electrons, and CO needs two electrons (Wu and Huang 2017). As a result, different electron concentration on different reaction sites can lead to different reaction pathways and products.

Metal single atom usually have an affinity for CO_2 molecules; various single atoms have been doped into the catalyst to form the active sites for photocatalytic CO_2 reduction. For instance, Gao et al. (Gao et al. 2016) have studied the influence of platinum and palladium single atom on the photocatalytic performance of graphite carbon nitride (g-C_3N_4). The platinum and palladium single atom tend to dope into the graphite carbon with binding energy of 2.95 and 2.17 eV, respectively. Notably, platinum and palladium single atom leads to different final product. The product of Pd/g-C_3N_4 is HCOOH, which has an energy barrier of 0.66 eV. The reaction energy barrier for Pt/g-C_3N_4 is 1.16 eV, which is benefit for producing the CH_4 product. The different final product may be caused by different affinity with hydrocarbon reactant, which led to the different energy barrier and the different product.

Liu et al. (Liu et al. 2018) load Au clusters on TiO_2 nanoflake (001) surface, which has rich O vacancy and Ti^{3+}. The chemical adsorption energy of CO_2 changed from 0.25 eV to 1.87 eV after loading with Au clusters. In addition, the Au cluster promotes the conversion of CO to CH$_4$. In short, single-transition metal atom with empty track is conducive to CO$_2$ adsorption and activation, which can promote the combination with the reactants. In general, according to the first principle, the d band center of transition metal is close to the Fermi level and in favor of CO$_2$ activation. DFT calculation shows that (Zhao, Chen, and Yang 2019), compared with Pt nanoparticles, the single platinum atom on Ti$_3$C$_2$ exhibited partial positive charge and atomic dispersion characteristic, which is beneficial to reduce the adsorption energy and activation energy of CO$_2$, thus improving the performance of CO$_2$ reduction. For example, Bi ions absorbed on TiO$_2$ nanosheets for CO$_2$ photoreduction was studied (Liao et al. 2014). The separated double Bi ions promote the formation of an electric field, thus promoting the separation of photogenerated charge carriers. Moreover, TiO$_2$-Bi is more suitable for the CH$_4$ production.

Wang et al. (Zhang, Zhao, et al. 2018) prepared double-transition metal MXene (Mo$_2$TiC$_2$T$_x$) nanosheets with abundant surface Mo vacancies (V$_{Mo}$) by electrochemical exfoliation (Mo$_2$TiC$_2$T$_x$-V$_{Mo}$), in which the Mo vacancies can act as the anchoring sites for single Pt atoms (Mo$_2$TiC$_2$T$_x$-Pt$_{SA}$). During the electrochemical exfoliation process, single Pt atoms are simultaneously immobilized on the Mo vacancies and are stabilized by the formation of covalent Pt–C bonds with the surrounding C atoms on the MXene. The resultant Mo$_2$TiC$_2$T$_x$-Pt$_{SA}$ catalysts show excellent stability, due to the strong bonding strength between Pt atoms and Mo$_2$TiC$_2$T$_x$.

Different transition metal vacancies on different MXenes can trapped varied metal atoms. Liu et al. (Li, Zhang, et al. 2020) load Ru atom on Ti$_3$C$_2$ and in-situ oxidation of Ti$_3$C$_2$ to obtain the TiO$_2$-Ti$_3$C$_2$/Ru. The size of Ru nanoparticles was approximately 3 nm on the Ti$_3$C$_2$ surfaces of the Ti$_3$C$_2$/Ru sample (Figure 7.9). In the TiO$_2$-Ti$_3$C$_2$/Ru system, the Ti$_3$C$_2$/Ru co-catalyst with higher work function

FIGURE 7.9 Transmission electron microscopy image of Ti$_3$C$_2$/Ru (Li, Zhang, et al. 2020).

reflecting the lower Fermi level, which greatly promotes the transfer rate of photogenerated electrons from TiO_2 to Ti_3C_2/Ru co-catalyst. In addition, the TiO_2-
Ti_3C_2/Cu and TiO_2-Ti_3C_2/Ni catalysts were also successfully prepared in the same
way. Therefore, we can stabilize appropriate single atom on the position of Ti defect
of MXene by in situ self-reduction process at room temperature, by forming the
strong metal-carbon bonds with the MXene carrier, the performance of the photocatalytic CO_2 reduction reaction can be significantly improved.

Bimetallic structures can be obtained by induces a chemical reaction between noble
metal and the support. The catalytic active sites can be efficiently tuned by the reaction
between the noble metal nanoparticles and MXenes (Li, Cui, et al. 2018). There is an
interaction between the platinum and Nb_2CT_x MXene after reducing at 350 °C in H_2
atmosphere. The interaction of Pt and MXenes is also observed in $Pt/Ti_3C_2T_x$, and the
bimetallic interface is beneficial for preparing high efficient catalysts.

7.4.4 THEORETICAL CALCULATION

Density functional theory (DFT) calculation is an approximate method to predict
the property and performance of the photocatalysts. The essence of the DFT method
is to use the electron density as the carrier of all information in the ground state of
the molecule or atom, rather than the wave function of a single electron, so that the
multi-electron system is transformed into a single-electron problem. DFT calculation have been widely used in the field of photocatalytic CO_2 reduction (Yang et al.
2021). For example, the adsorption and activation of CO_2, and charge transfer
mechanism can be investigated by the DFT calculation. In this section, the DFT
calculation for the MXene-based photocatalyst will be discussed.

7.4.4.1 MXene with Defect

During the etching process of Ti_3AlC_2 to obtain the Ti_3C_2 MXene, some point defects
will appear on the surface of Ti_3C_2 (Benchakar and Loupias 2020). By adjusting the
concentration of etchant, the concentration of point defects can be changed. For instance, when the Ti atoms around the defect sites changed from three-O-coordinated
into two-coordinated, some free Ti sites were generated on the surface of MXene to
improve the catalytic activity. The defects have an influence on the surface morphology and terminal groups, but they have no significant influence on the electrical
conductivity of the metal (McConnell, Li, and Brudvig 2010).

Numerous reports have shown that oxygen vacancy (O_v) is the active site for
various reaction (Shi and Chen 2019). Under high temperature, the surface of –OH
groups can be converted into –O groups (Persson et al. 2017). Oxygen vacancy at
Ti_2CO_2, V_2CO_2, and $Ti_3C_2O_2$ is studied by DFT calculation to investigate the effect
of O_v on the catalytic activity for CO_2 reduction (Zhang et al. 2017). As is well-
known, the first step of CO_2 reaction is the adsorption of CO_2 on the active center of
the catalyst.

The adsorption energies between CO_2 molecules and O-terminated MXene with
O_v range from -0.35 eV to –0.73eV, indicating that O_v at MXene surface could
effectively capture the CO_2 around the active sites for subsequent reduction reaction. The adsorption of CO_2 at the O_v of Ti_2CO_2 and $Ti_3C_2O_2$ is stronger than that

of V$_2$CO$_2$ since Ti can easily lose electrons compared with V. In addition, it is worth noting that the other initial reactant H$_2$O is also affected by the oxygen vacancy on the surface. The adsorption of CO$_2$ and H$_2$O is competitive on the MXene surface, and the oxygen vacancy is not conducive to H$_2$O adsorption. Therefore, the production of H$_2$ is inhibited and the performance of photocatalytic CO$_2$ reduction is enhanced.

The density functional theory calculation suggest that the –O and –OH terminated MXene are beneficial for selective conversion of CO$_2$ to CH$_4$ due to the reduced overall reaction free energy compared to the bare MXene (Nguyen et al. 2020). Figure 7.10 displayed the minimum energy pathway for the reduction of CO$_2$ to CH$_4$ on the –OH and –O terminated Mo$_3$C$_2$, respectively. As shown in Figure 7.10a, the reaction energy for the first hydrogenation step (–CO$_2$ → HOCO*) is -0.92 eV, which is smaller than that of bare Mo$_3$C$_2$ (–0.64 eV), indicating the essential role of –OH groups in the hydrogenation of CO$_2$. Moreover, the energy release of the subsequent steps suggested the higher activity of Mo$_3$C$_2$(OH)$_2$ than bare Mo$_3$C$_2$. Similarly, it can be observed from Figure 7.10b that Mo$_2$C$_2$O$_2$ requires energy to form the *COOH, *CH$_2$O, and *CH$_3$OH radicals, while the rest steps are the exothermic processes. Therefore, it can be speculated that –OH and –O terminated Mo$_3$C$_2$ are promising candidates for the CO$_2$ reduction reaction.

Very recently, Sc$_2$C(OH)$_2$ and Y$_2$C(OH)$_2$ MXenes containing –OH termination groups were explored as feasible catalysts for the CO$_2$ reduction reaction via first-principle simulation (Chen et al. 2019). It was found that Sc$_2$C(OH)$_2$ and Y$_2$C(OH)$_2$ MXenes possess less negative potential than Cu. Furthermore, the H atom in –OH serve an essential function to form the stable species with intermediates and decrease the overpotential. Consequently, efficiency of the CO$_2$ reduction reaction over the Sc$_2$C(OH)$_2$ and Y$_2$C(OH)$_2$ MXenes is greatly enhanced.

According to the DFT calculations, we can draw some conclusions. (1) The defects have an influence on the surface morphology and terminal groups, but they have no significant influence on the electrical conductivity of MXene. (2) Oxygen vacancy at the MXene surface can effectively capture the reactant around the active sites for subsequent catalytic reaction. (3) The surface terminal groups, such as –O and –OH groups, modified MXenes significantly impacts the reaction pathways. However, few experimental investigations of MXenes for photocatalytic CO$_2$ reduction was reported, and numerous future efforts are warranted to provide more experimental results, which can verify the outstanding properties predicted by the DFT calculations.

7.4.4.2 MXene with Single Atom

The key point to obtained the efficient single atom catalyst is to establish a strong interaction between the single atom and the support. Several reports highlight the important role of single atoms on MXene to increase the photocatalytic performance. For example, Zhang et al (Zhang, Zhao, et al. 2018) synthesized the Mo$_2$TiC$_2$T$_x$ nanosheets with a large number of exposed bases and outer layer Mo vacancies via electrochemical peeling. The formed Mo vacancies on Mo$_2$TiC$_2$T$_x$ can be used to fix the single Pt atoms (Mo$_2$TiC$_2$T$_x$-Pt$_{SA}$) (Figure 7.11), thus enhancing the catalytic hydrogen evolution activity of MXene. Notably, replacing Mo

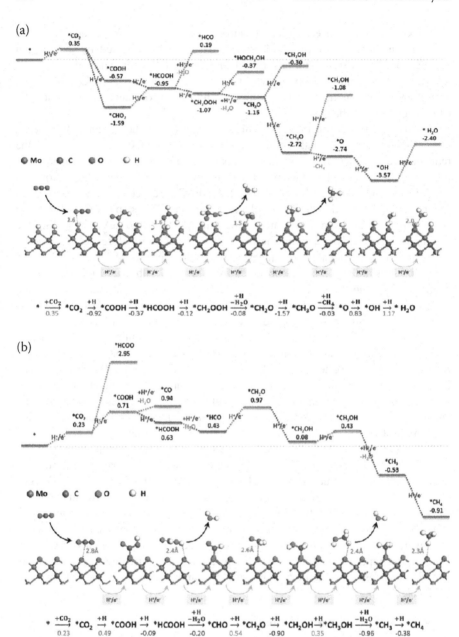

FIGURE 7.10 Minimum energy pathway for the conversion of CO_2 into CH_4 and H_2O on (a) $Mo_3C_2(OH)_2$ and (b) $Mo_3C_2O_2$ (Li, Chen, and Ong 2017). Figure 7.10 displayed the minimum energy pathway for the reduction of CO_2 to CH_4 on the –OH and –O terminated Mo_3C_2, respectively. As shown in Figure 7.10a, the reaction energy for the first hydrogenation step (–CO_2 → HOCO·) is –0.92 eV, which is smaller than that of bare Mo_3C_2 (–0.64 eV), indicating the essential role of –OH groups in the hydrogenation of CO_2. Moreover, the energy release of the subsequent steps suggested the higher activity of $Mo_3C_2(OH)_2$ than bare Mo_3C_2. Similarly, it can be observed from Figure 7.10b that $Mo_2C_2O_2$ requires energy to form the *COOH, *CH_2O, and *CH_3OH radicals, while the rest steps are the exothermic processes.

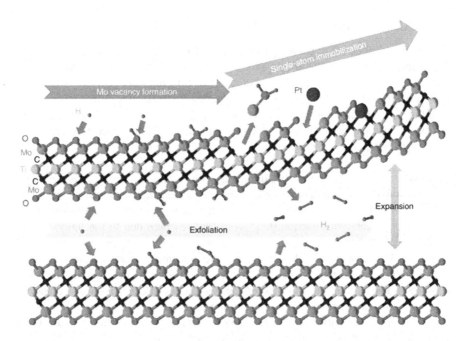

FIGURE 7.11 The synthesis mechanism for $Mo_2TiC_2O_2$–PtSA (Zhang, Zhao, et al. 2018). Mo_2TiC_2Tx nanosheets with a large number of exposed bases and outer layer Mo vacancies via electrochemical peeling. The formed Mo vacancies on $Mo_2TiC_2T_x$ can be used to fix the single Pt atoms ($Mo_2TiC_2T_x$-PtSA).

atom with the Pt atom can caused the decreased charge density around the Pt atom. In the projected density of states near the Fermi level, Pt single atom-immobilized MXene showed a higher occupied state than pure MXene. Moreover, the higher Fermi energy level of $Mo_2TiC_2T_x$-Pt_{SA} lead to a lower work function. The higher electronic energy level of $Mo_2TiC_2T_x$-Pt_{SA} would lead to a stronger ability to provide electrons. Therefore, catalytic ability can be greatly improved with the Pt single atoms combining to the MXene surface.

DFT calculation can also reveal the reaction mechanism and highlight the superior catalytic performance. Zhao et al. (Zhao, Chen, and Yang 2019) synthesized ultrathin $Ti_{3-x}C_2T_y$ nanosheets with stable Pt single atom via a simultaneous self-reduction stabilization process at room temperature. The $Ti_{3-x}C_2T_y$ nanosheets with abundant Ti defects exhibited high reduction ability. Therefore, the Pt atom can be stabilized in the Ti defects and form a strong metal-carbon bond with $Ti_{3-x}C_2T_y$. In this process, the catalytic performance was optimized due to the lower energy barrier of CO_2 insertion into the Pt–H bond, dispersed single atoms, and partial positively charged Pt.

Based on theoretical calculations, the performance of photocatalytic CO_2 reduction reaction can be optimized by stabilizing the single atoms on MXene. However, more attention should be paid to experimental research to further confirm the DFT calculation results.

7.5 CONCLUSION AND PERSPECTIVES

MXene-based photocatalysts for photocatalytic CO_2 reduction was summarized in this chapter. Due to the electronic conductivity and high surface area, MXene has been widely used as co-catalyst of photocatalyst for photocatalytic CO_2 reduction. However, the study of the MXene-based for photocatalytic CO_2 reduction is still in its infancy, far behind the practical application. Therefore, more effort is required for the experiment research and DFT calculation of MXene-based photocatalysts for further enhancing the performance of photocatalytic CO_2 reduction.

1. MXenes have been widely applied as co-catalysts of photocatalyst for photocatalytic CO_2 reduction. However, most researches mainly focused on Ti_3C_2 MXene, few reports can be found for the other MXene materials used for photocatalytic reaction. It is worth our research due to the unique properties of the other MXenes. Moreover, due to the two-dimensional structure of the MXene, more two-dimensional semiconductors should be discovered and combined with the MXene to form the 2D/2D compact interface. In this way, the transfer path and distance of the photoinduced electrons will be shorter and the electrons can rapidly transfer to the active sites on the surface. Therefore, formation of 2D/2D MXene/semiconductor heterostructure is an effective method for the design and synthesis of efficient photocatalysts.

2. On the one hand, Formation of the Z-scheme system is an effective way to improve the separation efficiency of photogenerated charge carriers. On the other hand, Schottky-junction can be formed by coupling with the MXene with semiconductors, which is conducive to the rapid transfer of electrons from the semiconductor to the MXene and significantly enhanced the separation of the photoinduced electrons and holes. Therefore, combining the MXene with the Z-scheme system might be a feasible way to further restrain the recombination of the photogenerated electrons and holes.

3. More attention should be paid to the stability of MXene. MXene is not very stable in the aqueous solution, it can be oxidized during the photocatalytic reaction when the reaction was carried out in the aqueous solution. However, the decomposition behavior and oxidation mechanism of MXene are still unclear. For example, termination groups on the surface of MXene is hard to quantify and control, and the termination groups have a great effect on the stability of the MXene. Therefore, to figure out the reason of the decomposition and oxidation of MXene is a matter of urgency.

4. The products obtained from photocatalytic CO_2 reduction is complicated, such as CH_4, H_2, CH_3OH, etc. There is still a long way to achieve the controlled selectivity of the products. Strategies, such as regulating redox potential, controlling active sites, interfacial bonds, and defects may be potential pathways to control the selectivity of photocatalytic CO_2 reduction.

In summary, MXenes, as a co-catalyst, has a broad application prospect in photocatalytic CO_2 reduction. However, more efforts still in need to obtained the highly stable and efficient MXene-based photocatalysts for photocatalytic CO_2 reduction.

ACKNOWLEDGMENTS

This work was supported by the National Natural Science Foundation of China (22078074, 21938001), Guangxi Natural Science Foundation (2019GXN-SFAA245006, 2020GXNSFDA297007, 2016GXNSFFA380015), Special funding for 'Guangxi Bagui Scholars', and Scientific Research Foundation for High-level Personnel from Guangxi University.

REFERENCES

Alhabeb, Mohamed, Kathleen Maleski, Babak Anasori, et al. 2017. Guidelines for synthesis and processing of two-dimensional titanium carbide (Ti$_3$C$_2$T$_x$ MXene). *Chemistry of Materials* 29 (18):7633–7644.

Alvarez, A., A. Bansode, A. Urakawa, et al. 2017. Challenges in the Greener Production of Formates/Formic Acid, Methanol, and DME by Heterogeneously Catalyzed CO$_2$ Hydrogenation Processes. *Chemical Reviews* 117 (14):9804–9838.

Anasori, Babak, Maria R. Lukatskaya, and Yury Gogotsi. 2017. 2D metal carbides and nitrides (MXenes) for energy storage. *Nature Reviews Materials* 2:16098.

Benchakar, Mohamed and Lola Loupias. 2020. One MAX phase, different MXenes: A guideline to understand the crucial role of etching conditions on Ti$_3$C$_2$T$_x$ surface chemistry. *Applied Surface Science* 530:147209.

Chen, Hetian, Albertus. D. Handoko, Jiwen Xiao, et al. 2019. Catalytic effect on CO$_2$ electroreduction by hydroxyl-terminated two-dimensional MXenes. *ACS Applied Materials & Interfaces* 11 (40):36571–36579.

Chen, Weiyi, Bin Han, Yili Xie, Shujie Liang, Hong Deng, and Zhang Lin. 2020. Ultrathin Co-Co LDHs nanosheets assembled vertically on MXene: 3D nanoarrays for boosted visible-light-driven CO$_2$ reduction. *Chemical Engineering Journal* 391:123519.

Dong, Hongjun, Xiaoxu Zhang, Yan Zuo, et al. 2020. 2D Ti$_3$C$_2$ as electron harvester anchors on 2D g-C$_3$N$_4$ to create boundary edge active sites for boosting photocatalytic performance. *Applied Catalysis A: General* 590:117367.

Fu, Zhiyan, Qi Yang, Zhan Liu, et al. 2019. Photocatalytic conversion of carbon dioxide: From products to design the catalysts. *Journal of CO$_2$ Utilization* 34:63–73.

Gao, Guoping, Yan Jiao, Eric. R. Waclawik, et al. 2016. Single atom (Pd/Pt) supported on graphitic carbon nitride as an efficient photocatalyst for visible-light reduction of carbon dioxide. *Journal of the American Chemical Society* 138 (19):6292–6297.

Ghidiu, M., M. Naguib, C. Shi, et al. 2014. Synthesis and characterization of two-dimensional Nb$_4$C$_3$ (MXene). *Chemical Communications* 50 (67):9517–9520.

Giordano, Cristina, Christian Erpen, Weitang Yao, Bettina Milke, and Markus Antonietti. 2009. Metal nitride and metal carbide nanoparticles by a soft urea pathway. *Chemistry of Materials* 21 (21):5136–5144.

Guo, Zhonglu, Ying Li, Baisheng Sa, et al. 2020. M$_2$C-type MXenes: Promising catalysts for CO$_2$ capture and reduction. *Applied Surface Science* 521:146436.

Han, Na, Yu Wang, Hui Yang, et al. 2018. Ultrathin bismuth nanosheets from in situ topotactic transformation for selective electrocatalytic CO$_2$ reduction to formate. *Nature Communications* 9 (1):1320.

He, Fei, Bicheng Zhu, Bei Cheng, Jiaguo Yu, Wingkei Ho, and Wojciech Macyk. 2020. 2D/2D/0D TiO$_2$/C$_3$N$_4$/Ti$_3$C$_2$ MXene composite S-scheme photocatalyst with enhanced CO$_2$ reduction activity. *Applied Catalysis B: Environmental* 272:119006.

Hong, Jindui, Wei Zhang, Jia Ren, and Rong Xu. 2013. Photocatalytic reduction of CO$_2$: A brief review on product analysis and systematic methods. *Analytical Methods* 5 (5):1073–1356.

Houghton, R. A. 2005. Aboveground forest biomass and the global carbon balance. *Global Change Biology* 11 (6):945–958.

Huang, Guimei, Shuangzhi Li, Lijun Liu, Leifan Zhu, and Qiang Wang. 2020. Ti_3C_2 MXene-modified Bi_2WO_6 nanoplates for efficient photodegradation of volatile organic compounds. *Applied Surface Science* 503:144183.

Intikhab, Saad, Varun Natu, Justin Li, et al. 2019. Stoichiometry and surface structure dependence of hydrogen evolution reaction activity and stability of Mo_xC MXenes. *Journal of Catalysis* 371:325–332.

Jiang, Yanan, Sun Tao, and Xie Xi. 2019. Oxygen functionalized ultrathin $Ti_3C_2T_x$ MXene for enhanced electrocatalytic hydrogen evolution. *ChemSusChem* 12 (7):1368–1373.

Jin-Shui, Zhang, Wang Bo, and Wang Xin-Chen. 2013. Chemical synthesis and applications of graphitic carbon nitride. *Acta Physico-Chimica Sinica* 29 (9):1865–1876.

Johnson, Magnus T. and Ola F. Wendt. 2014. Carboxylation reactions involving carbon dioxide insertion into palladium–carbon σ-bonds. *Journal of Organometallic Chemistry* 751:213–220.

Khaledialidusti, R., A. K. Mishra, and A. Barnoush. 2019. CO_2 adsorption and activation on the (110) chalcopyrite surfaces: A dispersion-corrected DFT + U study. *ACS Omega* 4 (14):15935–15946.

Khaledialidusti, Rasoul, Abhishek Kumar Mishra, and Afrooz Barnoush. 2020. Atomic defects in monolayer ordered double transition metal carbide ($Mo_2TiC_2T_x$) MXene and CO_2 adsorption. *Journal of Materials Chemistry C* 8 (14):4771–4779.

Khezri, Bahareh, Adrian C. Fisher, and Martin Pumera. 2017. CO_2 reduction: The quest for electrocatalytic materials. *Journal of Materials Chemistry A* 5 (18):8230–8246.

Li, Ang, Tuo Wang, Xiaoxia Chang, et al. 2018. Tunable syngas production from photocatalytic CO_2 reduction with mitigated charge recombination driven by spatially separated cocatalysts. *Chemical Science* 9 (24):5334–5340.

Li, Jinmao, Li Zhao, Shimin Wang, Jin Li, Guohong Wang, and Juan Wang. 2020. In situ fabrication of 2D/3D g-C_3N_4/Ti_3C_2 (MXene) heterojunction for efficient visible-light photocatalytic hydrogen evolution. *Applied Surface Science* 515:145922.

Li, Mingsai, Linxin, Zhang, et al. 2020. Regulating electron-hole separation to promote photocatalytic H_2 evolution activity of nanoconfined Ru/MXene/TiO_2 catalysts. *ACS Nano* 14 (10):14181–14189.

Li, Zhe, Yanran Cui, et al. 2018. Reactive metal–support interactions at moderate temperature in two-dimensional niobium-carbide-supported platinum catalysts. *Nature Catalysis* 1:349–355.

Liao, Yusen, Shao-Wen , et al. 2014. Efficient CO_2 capture and photoreduction by amine-functionalized TiO_2. *Chemistry – A European Journal* 20 (33):10220–10222.

Lima-Melo, Yugo, Peter J. Gollan, Mikko Tikkanen, et al. 2019. Consequences of photosystem-I damage and repair on photosynthesis and carbon use in Arabidopsis thaliana. *The Plant Journal* 97(6):1061–1072.

Liu, Lianjun, Huilei Zhao, Jean M. Andino, and Ying Li. 2012. Photocatalytic CO_2 reduction with H_2O on TiO_2 nanocrystals: Comparison of anatase, rutile, and brookite polymorphs and exploration of surface chemistry. *ACS Catalysis* 2 (8):1817–1828.

Liu, Xueyan, Meng Ye, Shuping Zhang, et al. 2018. Enhanced photocatalytic CO_2 valorization over TiO_2 hollow microspheres by synergetic surface tailoring and Au decoration. *Journal of Materials Chemistry A* 6 (47):24245–24255.

Low, Jingxiang, Liuyang Zhang, Tong Tong, Baojia Shen, and Jiaguo Yu. 2018. TiO_2/MXene Ti_3C_2 composite with excellent photocatalytic CO_2 reduction activity. *Journal of Catalysis* 361:255–266.

McConnell, I., G. Li, and G. W. Brudvig. 2010. Energy conversion in natural and artificial photosynthesis. *Chemistry & Biology* 17 (5):434–447.

Meng, A., L. Zhang, B. Cheng, and J. Yu. 2019a. Dual cocatalysts in TiO$_2$ photocatalysis. *Advanced Materials* 31 (30):1807660.

Meng, A., L. Zhang, B. Cheng, and J. Yu . 2019b. TiO$_2$-MnO$_x$-Pt hybrid multiheterojunction film photocatalyst with enhanced photocatalytic CO$_2$-reduction activity. *ACS Applied Materials & Interfaces* 11 (6):5581–5589.

Morales-Garcia, A., M. Mayans-Llorach, F. Vines, and F. Illas. 2019. Thickness biased capture of CO$_2$ on carbide MXenes. *Physical Chemistry Chemical Physics* 21 (41):23136–23142.

Morales-García, Ángel, Adrián Fernández-Fernández, Francesc Viñes, and Francesc Illas. 2018. CO$_2$ abatement using two-dimensional MXene carbides. *Journal of Materials Chemistry A* 6 (8):3381–3385.

Natu Varun, Maxim Sokol, and Louisiane Verger. 2018. Effect of edge charges on stability and aggregation of Ti$_3$C$_2$T$_z$ MXene colloidal suspensions. *Journal of Physical Chemistry C* 122:27745–27753.

Neng, Neng, Xingzhu Chen, and Wee-Jun Ong. 2017. Understanding of electrochemical mechanisms for CO$_2$ capture and conversion into hydrocarbon fuels in transition-metal carbides (MXenes). *ACS Nano* 11 (11):10825–10833.

Nguyen, Thang Phan, Dinh Minh Tuan Nguyen, Dai Lam Tran, et al. 2020. MXenes: Applications in electrocatalytic, photocatalytic hydrogen evolution reaction and CO$_2$ reduction. *Molecular Catalysis* 486:110850.

Nocera, Nathan S. and Lewis, Daniel G. 2007. Powering the planet: Chemical challenges in solar energy utilization. *Proceeding of the National Academy of Science USA* 104 (42):15729–15735.

Park, Jae Hyun, Jeongwoo Yang, Dohyeun Kim, Hyeonseo Gim, Won Yeong Choi, and Jae W. Lee. 2022. Review of recent technologies for transforming carbon dioxide to carbon materials. *Chemical Engineering Journal* 427:130980.

Peng, Jiahe, Xingzhu Chen, Wee-Jun Ong, Xiujian Zhao, and Neng Li. 2019. Surface and heterointerface engineering of 2D MXenes and their nanocomposites: Insights into electro- and photocatalysis. *Chem* 5 (1):18–50.

Persson, Ingemar, Lars-Åke Näslund, Joseph Halim, et al. 2017. On the organization and thermal behavior of functional groups on Ti$_3$C$_2$ MXene surfaces in vacuum. *2D Materials* 5 (1):015002.

Qu, Di, Xianyun Peng, et al. 2020. Nitrogen doping and titanium vacancies synergistically promote CO$_2$ fixation in seawater. *Nanoscale* 12 (33):17191–17195.

Qu, Guoxing, Yang Zhou, and Tianli Wu, et al. 2018. Phosphorized MXene-phase molybdenum carbide as an earth-abundant hydrogen evolution electrocatalyst. *ACS Applied Energy Materials* 1 (12):7206–7212.

Raizada, Pankaj, Anita Sudhaik, Pardeep Singh, Ahmad Hosseini-Bandegharaei, and Pankaj Thakur. 2019. Converting type II AgBr/VO into ternary Z scheme photocatalyst via coupling with phosphorus doped g-C3N$_4$ for enhanced photocatalytic activity. *Separation and Purification Technology* 227:115692.

Seh, Zhi Wei, Kurt D. Fredrickson, Babak Anasori, et al. 2016. Two-dimensional molybdenum carbide (MXene) as an efficient electrocatalyst for hydrogen evolution. *ACS Energy Letters* 1 (3):589–594.

Shi, Rui and Yong Chen. 2019. Controlled formation of defective shell on TiO$_2$(001) facets for enhanced photocatalytic CO$_2$ reduction. *ChemCatChem* 11 (9):2270–2276.

Tahir, Muhammad. 2020. Construction of a stable two-dimensional MAX supported protonated graphitic carbon nitride (pg-C$_3$N$_4$)/Ti$_3$AlC$_2$/TiO$_2$ Z-scheme multiheterojunction system for efficient photocatalytic CO$_2$ reduction through dry reforming of methanol. *Energy & Fuels* 34 (3):3540–3556.

Tang, Qijun and Zhuxing Sun. 2020. Decorating g-C$_3$N$_4$ with alkalinized Ti$_3$C$_2$ MXene for promoted photocatalytic CO$_2$ reduction performance. *Journal of Colloid Interface Science* 564:406–417.

Tang, Yi, Chenhui Yang, Yapeng Tian, Yangyang Luo, Xingtian Yin, and Wenxiu Que. 2020. The effect of in situ nitrogen doping on the oxygen evolution reaction of MXenes. *Nanoscale Advances* 2 (3):1187–1194.

Teramura, K. and T. Tanaka. 2018. Necessary and sufficient conditions for the successful three-phase photocatalytic reduction of CO_2 by H_2O over heterogeneous photocatalysts. *Physical Chemistry Chemical Physics* 20 (13):1187–1194.

Tu, Wenguang, You Xu, Jiajia Wang, et al. 2017. Investigating the role of tunable nitrogen vacancies in graphitic carbon nitride nanosheets for efficient visible-light-driven h_2 evolution and CO_2 reduction. *ACS Sustainable Chemistry & Engineering* 5 (8):7260–7268.

Wan, Yan-Jun, Krishnamoorthy Rajavel, and Xing-Miao Li. 2020. Electromagnetic interference shielding of $Ti_3C_2T_x$ MXene modified by ionic liquid for high chemical stability and excellent mechanical strength. *Chemical Engineering Journal* 408:127303.

Wang, Bingxin, Aiguo Zhou, Fanfan Liu, Jianliang Cao, Libo Wang, and Qianku Hu. 2018. Carbon dioxide adsorption of two-dimensional carbide MXenes. *Journal of Advanced Ceramics* 7 (3):237–245.

Wang, Yanan, Wenlong Zhen, Yiqing Zeng, et al. 2020. In situ self-assembly of zirconium metal–organic frameworks onto ultrathin carbon nitride for enhanced visible light-driven conversion of CO_2 to CO. *Journal of Materials Chemistry A* 8 (12):6034–6040.

Wang, Ying and Zizhong Zhang. 2018. Visible-light driven overall conversion of CO_2 and H_2O to CH_4 and O_2 on 3D-SiC@2D-MoS_2 heterostructure. *Journal of the American Chemical Society* 140 (44):14595–14598.

White, James L., Maor F. Baruch, et al. 2015. Light-driven heterogeneous reduction of carbon dioxide: Photocatalysts and photoelectrodes. *Chemical Reviews* 115:12888–12935.

Wu, Jinhua and Yang Huang. 2017. CO_2 Reduction: From the electrochemical to photochemical approach. *Advanced Science* 4:1700194–1700213.

Xu, Yue, Yong You, et al. 2019. Bi_4NbO_8Cl {001} nanosheets coupled with g-C_3N_4 as 2D/2D heterojunction for photocatalytic degradation and CO_2 reduction. *Journal of Hazardous Materials* 381 (12):121159.

Yang, Chao, Qiuyan Tan, Qin Li, et al. 2020. 2D/2D Ti_3C_2 MXene/g-C_3N_4 nanosheets heterojunction for high efficient CO_2 reduction photocatalyst: Dual effects of urea. *Applied Catalysis B: Environmental* 268:118738.

Yang, Jun, Quanxi Zhu, Zhigang Xie, et al. 2021. Enhancement mechanism of photocatalytic activity for MoS_2/Ti_3C_2 Schottky junction: Experiment and DFT calculation. *Journal of Alloys and Compounds* 887:161411.

Yoon, Yeoheung, Anand P. Tiwari, Minhe Lee, et al. 2018. Enhanced electrocatalytic activity by chemical nitridation of two-dimensional titanium carbide MXene for hydrogen evolution. *Journal of Materials Chemistry A* 6 (42):20869–20877.

Yu, Huijun, Shi Run, Yunxuan Zhao, et al. 2017. Alkali-assisted synthesis of nitrogen deficient graphitic carbon nitride with tunable band structures for efficient visible-light-driven hydrogen evolution. *Advanced Materials* 29 (16):1605148.

Yuan, Lan, Sung-Fu Hung, Zi-Rong Tang, Hao Ming Chen, Yujie Xiong, and Yi-Jun Xu. 2019. Dynamic evolution of atomically dispersed Cu species for CO_2 photoreduction to solar fuels. *ACS Catalysis* 9 (6):4824–4833.

Zeng, Zhiping, Yibo Yan, Jie Chen, Ping Zan, Qinghua Tian, and Peng Chen. 2019. Boosting the photocatalytic ability of Cu_2O nanowires for CO_2 conversion by MXene quantum dots. *Advanced Functional Materials* 29 (2):1806500.

Zhang, Jiangqing, Yufei Zhao, et al. 2018. Single platinum atoms immobilized on an MXene as an efficient catalyst for the hydrogen evolution reaction. *Nature Catalysis* 1:985–992.

Zhang, Ruiyang, Huang Zeai, and Chengjing Li. 2019. Monolithic g-C_3N_4/reduced graphene oxide aerogel with in situ embedding of Pd nanoparticles for hydrogenation of CO_2 to CH_4. *Applied Surface Science* 475:953–960.

Zhang, Shijie, Han Zhuo, Suiqin Li, et al. 2020. Effects of surface functionalization of mxene-based nanocatalysts on hydrogen evolution reaction performance. *Catalysis Today* 368:187–195.

Zhang, Wenjun, Yi Hu, et al. 2018. Progress and perspective of electrocatalytic CO_2 reduction for renewable carbonaceous fuels and chemicals. *Advanced Science* 5 (1):1700275.

Zhang, Xu, Zihe Zhang, Jielan Li, Xudong Zhao, Dihua Wu, and Zhen Zhou. 2017. Ti_2CO_2 MXene: A highly active and selective photocatalyst for CO_2 reduction. *Journal of Materials Chemistry A* 5 (25):12899–12903.

Zhao, Di, Zheng Chen, and Wenjun Yang. 2019. MXene (Ti_3C_2) Vacancy-Confined Single-Atom Catalyst for Efficient Functionalization of CO_2. *Journal of American Chemical Society* 141 (9):4086–4093.

Zhao, Fengnian and Yao Yao. 2020. Self-reduction bimetallic nanoparticles on ultrathin MXene nanosheets as functional platform for pesticide sensing. *Journal of Hazardous Materials* 384:121358.

Zhao, Jia-Hui, Ling-Wang Liu, Kui Li, Tao Li, and Fu-Tian Liu. 2019. Conductive Ti_3C_2 and MOF-derived CoS_x boosting the photocatalytic hydrogen production activity of TiO_2. *CrystEngComm* 21 (14):2416–2421.

8 Application of MXene-Based Photocatalyst for Photocatalytic Degradation

Chengzheng Men and Tongming Su
School of Chemistry and Chemical Engineering, Guangxi University

CONTENTS

8.1 INTRODUCTION

The rapid development of industrialization has led to the massive discharge of hazardous waste, which is very harmful to the living environment of humans (Akhade et al. 2020; Wang, Ren, et al. 2020). Hazardous wastes, such as benzene, antioxidants, sulfonamides and tetracycline hydrochloride in cosmetics, methylene blue and rhodamine B in clothing dyes, sulfa antibiotics, carbamazepine, endocrine disruptors in the pharmaceutical industry, and pesticides, pose a threat to ecosystems and humans (Cheng et al. 2019; Huang, Li, et al. 2020; Kumar, Kumar, and Krishnan 2020; Rojas and Horcajada 2020; Tang, Xiong, et al. 2020; Wang, Jones, and Zhang 2020). Therefore, finding a suitable degradation strategy for organic pollution has become a major challenge in the treatment of organic pollutants in wastewater. Most organic pollutants can be removed by biodegradation, chemical oxidation, electrochemical conversion, and combustion (Naguib et al. 2011; Li et al. 2019; Huang, Pradhan, et al. 2020; Khan et al. 2020; Xie and Zhang 2020; Zhao

DOI: 10.1201/9781003156963-8

et al. 2020). However, some harmful organic pollutants are difficult to eliminate using the above technologies. Photocatalytic technology is an advanced oxidation technology with the advantages of being highly efficient, environmentally friendly, and low cost, and it has shown great potential in the purification of water. By using the photocatalytic technology, the organic pollutants can be converted into CO_2 and H_2O. Therefore, various semiconductor photocatalysts were discovered for photo-catalytic degradation of pollutants during the past decades. Photocatalytic degradation of harmful organic pollutants has attracted more and more attention.

The development of high-efficiency heterojunction of composite photocatalyst is a feasible way to improve the overall efficiency of photocatalytic reactions. In 2011, MXene, a new two-dimensional (2D) transition metal carbides, nitrides, and car-bonitrides were discovered (Naguib et al. 2011). Due to its excellent electronic conductivity and 2D structure, MXene has received more and more attention in the field of energy and environment. For example, MXene was widely studied in the field of ion batteries, catalysts, biomedicine, and energy storage (Peng et al. 2019). In addition, due to the high work function and electronic conductivity of the MXene (Zhao et al. 2015; Zhang et al. 2017; Xiao et al. 2019; Tang, Zhou, et al. 2020; Yun et al. 2020; Zhang et al. 2021), MXene has been used as a co-catalyst of photo-catalyst for photocatalytic hydrogen evolution, carbon dioxide reduction, pollutant degradation, and nitrogen fixation.

In the MXene/semiconductor heterojunction system, the formed interface between the MXene and semiconductor and the Schottky heterojunction can accelerate the charge-carrier separation, and enhance photocatalytic performance. Moreover, the large specific surface area of MXene can provide more active sites for photocatalytic redox reactions, which can affect the performance and selectivity of the photocatalyst. Therefore, MXene-based photocatalysts have attracted significant attention in the field of photocatalysis (Zhao et al. 2017; Khazaei et al. 2019; Kong et al. 2020; Li, Li, et al. 2020). The role of MXene in the MXene-based photocatalytic system is mainly considered as a photogenerated electron acceptor because it has a lower Fermi level than most semiconductor photocatalysts. However, due to its large specific surface area and abundant surface functional groups, the function of MXene in improving photocatalytic activity is not limited to photogenerated electron acceptors. MXenes have been used as co-catalysts for various semiconductor photocatalysts, such as ZnS, CdS, TiO_2, g-C_3N_4, TiO_2, ZnO, $BiWO_4$, $BiVO_4$, and $ZnIn_2S_4$ (Peng et al. 2019; Qin et al. 2019; Rasool et al. 2019; Ronchi, Arantes, and Santos 2019; Sinopoli et al. 2019; Nguyen et al. 2020; Prasad et al. 2020).

As is well-known, the rapid recombination of photogenerated charge carriers will severely limit the photocatalytic activity of a single photocatalyst. Therefore, using a promoter to modify the photocatalyst and separate photogenerated carriers is one of the most effective ways to improve photocatalytic activity. When MXene are combined with a semiconductor, a Schottky barrier can be constructed at the heterojunction interface, and this electronic barrier as an electronic container, thus facilitating the separation of photogenerated electrons and holes. According to density functional theory (DFT) calculation, the conductivity and Fermi level of MXene are lower than most n-type semiconductors, indicating that MXene can be used as a cocatalyst and act as a photo-generated electron transmission carrier and reactive site (Ran et al. 2017).

Moreover, the energy band of MXene can be controlled, such as controlling the lowest unoccupied molecular orbital (LUMO) and highest occupied molecular orbital (HOMO). Taking the photocatalytic degradation of wastewater as an example, because Ti_3C_2 has good electrical conductivity and suitable H^+ adsorption free energy (ΔG_H), H^+ can be effectively adsorbed on the surface of MXene and reduced to H_2, and a large number of photo-generated holes with strong oxidation ability can be used to degrade organic pollutant in water (Ran et al. 2017). In some photocatalytic systems, these electrons transferred to the MXene surface can also be used to reduce carbon organics into hydrocarbon compounds or generate reactive oxygen radicals to degrade pollutants (Huang, Li, et al. 2020; Khan et al. 2020).

In summary, in the field of photocatalysis, MXene plays an important role in inhibiting the recombination of photogenerated charge carriers and providing redox reactive sites for photocatalytic reaction, thus greatly improving the photocatalytic performance of the photocatalyst.

This chapter summarizes the design, synthesis, and mechanism of the MXene-based photocatalyst for the degradation of environmental organic pollutants. For example, the methods to improve the photocatalytic performance of MXenes are discussed. The surface performance and catalyst performance of various MXenes-based catalysts are compared, and the current difficulties and application prospects are analyzed. This chapter will further explore the development of new MXenes as co-catalysts suitable for photocatalytic degradation of semiconductor photocatalysts and will provide some positive effects.

Three types of MXenes-based photocatalysts for photocatalytic degradation are summarized in this chapter: (1) binary heterostructure based on Schottky barrier; (2) multiple heterogeneity of multi-level electron and hole transfer paths Quality structure; (3) modified and modified heterostructure. This chapter will systematically summarize and analyze various methods and strategies based on MXenes-based heterojunction and Schottky junction. These structures can enhance the absorption of light, accelerate the separation and transfer of charges, and extend the life of carriers. We will systematically analyze the influence of the surface morphology, electronic structure, and catalytic activity of various semiconductor couplings on the specific application of photocatalytic degradation from the perspective of mechanism. Finally, practical suggestions are put forward for the design of Mxene-based photocatalysts suitable for degradation, and some enlightening prospects are put forward for related research in recent years.

8.2   BINARY MXENE-BASED PHOTOCATALYST

Schottky junction was widely studied for the heterostructure photocatalyst. Since the work function of a general metal is greater than that of a semiconductor, electrons can easily migrate from the conduction band of the semiconductor to the metal. In this way, electrons and holes are placed on different sides to form a built-in electric field between the semiconductor and metal, which effectively promotes the separation of photoexcited charge carriers. Precious metals, such as Pt, Au, and Ag, are usually used to construct Schottky junction with the semiconductor (Hisatomi and Domen 2019). On the one hand, these metals have a large work function, which is conducive to the migration of photoelectrons from semiconductor to metal. On the other hand, precious metals are relatively stable during the photocatalytic reaction (Ji et al. 2005; Vitale, Curioni, and Andreoni 2008). In addition to precious metals, MXene also possesses a large surface work function, which indicates that it can be used as the electron accepter in the binary heterostructure photocatalyst. Generally, the work function of MXene has a strong dependence on its surface termination. As shown in Figure 8.1, the work function of MXene can be changed by adjusting the surface groups. For example, the work function of most MXenes with the -O termination exceeds that of Pt, which indicated that the noble metal Pt can be replace by MXenes in some cases (Liu, Xiao, and Goddard 2016). Moreover, the 2D ultrathin structure of MXene can provide numerous channels for electron transfer and more active sites for the photocatalytic reaction, which is also the advantage over precious metals.

Binary MXene-based photocatalysts have been widely used for photodegradation reaction. Their photodegradation mechanisms are basically similar, that is, the process of oxidation and removal of organic matter is completed in the valence band of the composite semiconductor, and the photogenerated electrons flow to MXenes through the contact interface. At the same time, the photoreduction process is completed on the surface of MXenes, which act as the electron accepter. Generally, in situ growth of a semiconductor on the MXene surface is an effective method to construct a binary MXene-based photocatalyst. It only needs to mix the precursor of the synthetic semiconductor with MXene, and then carry out the reaction at under certain conditions, such as appropriate temperature, pressure, pH, etc., to grow the semiconductor on the MXene. In this process, MXenes acts the role of the substrate for the semiconductor, and a compact interface can usually be formed between the MXene and the photocatalyst, which is beneficial for the separation of the photogenerated electrons

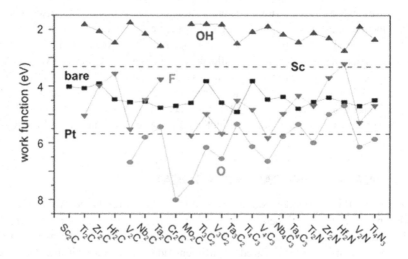

FIGURE 8.1 (a) Work functions of MXenes with various terminations. Bare surface, black square; O termination, red circle; OH, blue up-triangle; F, cyan down-triangle. For comparison, work functions of Sc and Pt metal are indicated by dashed lines. (b) Atomic structure of a representative M_2XT_2. M, purple; X, gray; T, red (Liu, Xiao, and Goddard 2016).

and holes. However, the active sites on the surface of the MXene might be cover by the semiconductor by using the in situ growth method.

The most prevalent in situ growth method to prepare the MXene-based photocatalyst is the one step of hydrothermal or solvothermal. For example, with $Ti_3C_2(O, OH)_x$ as the 2D platform for the Zn, In and S source, flower-like $Zn_2In_2S_5$ microspheres was produced in situ on the Ti_3C_2 under anaerobic hydrothermal conditions to obtain the $Zn_2In_2S_5/Ti_3C_2(O, OH)_x$ composite (Wang et al. 2019). The zeta potential of the $Ti_3C_2T_x$ was measured to be −28.6 mV due to the functional groups (−OH, −O, −F) on its surface. Therefore, before the temperature of the mixed precursor was heated up in this synthetic process, the layered $Ti_3C_2T_x$ can fix In^{3+} and Zn^{2+} cation in the aqueous solution due to the hydrophilic and negative charge surface of $Ti_3C_2T_x$. Moreover, with cetyltrimethylammonium bromide (CTAB) acting as the pre-pressing agent for In^{3+} and Zn^{2+} cation intercalation and surfactant, and the interlayer spacing of $Ti_3C_2T_x$ was increased through ion exchange, the uniform growth of $Zn_2In_2S_5$ can be controlled. With $Zn_2In_2S_5/Ti_3C_2(O, OH)_x$ as the photocatalyst, the removal rate of the tetracycline is 1.25 times that of pure $Zn_2In_2S_5$ under visible light irradiation. Furthermore, with the increase of temperature in the range of 35−55 °C, the removal rate of tetracycline can be further improved. The excellent photocatalytic activity originates from the synergistic effect between the visible light responsive $Zn_2In_2S_5$ and the conductive $Ti_3C_2(O, OH)_x$. In this system, the photoelectron transfer efficiency from $Zn_2In_2S_5$ to $Ti_3C_2(O, OH)_x$ is 33.0%, and the degradation of tetracycline mainly depends on the oxidation products of free radicals $^\bullet O^{2-}$ or h^+.

Bi_2MoO_6/Ti_3C_2 was also prepared and used for molecular oxygen activation (Li, Liu, et al. 2020). In this work, Ti_3C_2 was used as the 2D platform, and 2D ultrathin Bi_2MoO_6 nanosheets were grown in situ by the anaerobic hydrothermal method. In this way, 2D/2D Schottky heterojunction was formed between the Bi_2MoO_6 and Ti_3C_2, and the performance for molecular oxygen activation was 5.56 times higher than the Bi_2MoO_6. On the one hand, the large 2D/2D Schottky interface accelerated the separation of photogenerated electron-hole pairs. On the other hand, the high redox activity of the Ti sites on the surface of Ti_3C_2 promoted the multi-electron reduction reaction and achieves high performance of molecular-oxygen activation. During the photocatalytic reaction, Bi_2MoO_6 is excited by the incident light to generate electrons and holes. The photogenerated electrons first flow to the conduction band of Bi_2MoO_6, and then quickly transfer to Ti_3C_2 through the built-in electric field for reaction. The migrated photogenerated electrons can activate the absorbed molecular oxygen on the Ti_3C_2 nanosheets to generate free radicals $^\bullet O^{2-}$. Notably, because the Bi_2MoO_6 has a more negative valence band (2.35 eV < 2.40 eV), it cannot generate free radicals $^\bullet OH$ from the oxidation of H_2O by photogenerated holes. However, free radicals $^\bullet OH$ were detected in this system. Therefore, in this system, molecular oxygen was first activated to generate free radicals $^\bullet O^{2-}$, and then converted into other reactive oxygen species. The reactive oxygen species, including radical $^\bullet O^{2-}$, radical $^\bullet OH$ and $^1O_2$, generated by molecular oxygen activation, exhibited strong oxidizing ability and can participate in the photocatalytic degradation of organic pollutants (Figure 8.2).

MXene can simultaneously overcome the high charge carrier recombination rate and the serious photocorrosion problem of photocatalyst. Cai et al. prepared the Ag_3PO_4/Ti_3C_2 composite photocatalyst and found that Ti_3C_2 can greatly improve

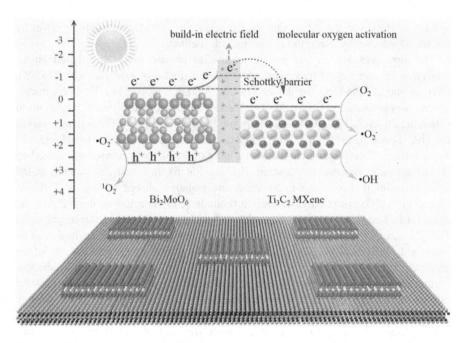

FIGURE 8.2 The possible mechanism for visible light driven molecular oxygen activation (Liu et al. 2020).

the photocatalytic activity and stability of Ag_3PO_4 (Cai et al. 2018). For this Ag_3PO_4/Ti_3C_2 photocatalyst, the abundant surface hydrophilic functional groups of Ti_3C_2 form a compact contact interface with Ag_3PO_4, which is beneficial to the separation of the photo-induced charge carriers. In addition, the surface Ti sites have strong redox activity. Moreover, the photogenerated electrons can transfer from the Ag_3PO_4 to the Ti_3C_2 surface rapidly through the built-in electric field, and the Schottky junction formed at the Ag_3PO_4/Ti_3C_2 interface inhibited the backflow of electrons. Therefore, Ag_3PO_4/Ti_3C_2 exhibits good photocatalytic activity and stability in degradation of organic pollutants. In particular, the apparent rate constant of Ag_3PO_4/Ti_3C_2 for the degradation of 2,4-dinitrophenol is 2.5 times and 10 times that of Ag_3PO_4/RGO and Ag_3PO_4, respectively. After eight cycles, the photocatalytic performance of Ag_3PO_4/Ti_3C_2 for tetracycline hydrochloride degradation remained at 68.4%, while the photocatalytic performance of $Ag_3PO_4/$ RGO and Ag_3PO_4 only maintained at 36.2% and 7.8%, respectively.

Ti_3C_2 can also be used as the cocatalyst of AgI for the effective photoreduction of Cr^{6+}, which indicates the great potential of Ti_3C_2 as the cocatalyst of various photosensitive semiconductors for improving their photocatalytic activity and stability (Sun et al. 2021). For example, MXene can be used as co-catalyst for the $CuFe_2O_4$ (Cao et al. 2020), CeO_2 (Shen et al. 2019), a-Fe_2O_3 (Zhang et al. 2018), and Bi_2WO_6 (Hu et al. 2020), photocatalyst.

It is worth mentioning that the synthesis conditions of the photocatalyst can directly affect the number of reactive sites. For example, vertical arrangement of MoS_2 was grown on the surface of Ti_3C_2 by hydrothermal method (Figure 8.3). The

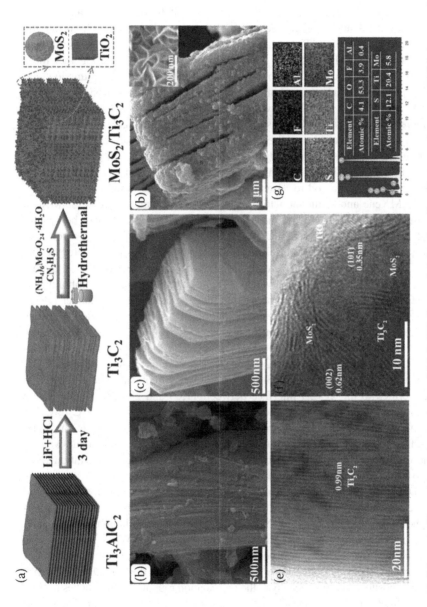

FIGURE 8.3 (a) Schematic illustration of the synthesis of $MoS_2@Ti_3C_2$ nanohybrid. Scanning electron microscopy (SEM) images of (b) Ti_3AlC_2; (c) Ti_3C_2, (d) $MoS_2@Ti_3C_2$ nanohybrid, and (inset in d) MoS_2 nanosheets on the surface of $MoS_2@Ti_3C_2$ nanohybrid. Transmission electron microscopy (TEM) images of (e) Ti_3C_2 and (f) $MoS_2@Ti_3C_2$ nanohybrid. (g) Energy-dispersive spectrum and mappings of $MoS_2@Ti_3C_2$ nanohybrid (Jiao and Liu 2019).

results show that friction is beneficial to improve the photocatalytic degradation efficiency because the shear force and pressure generated by the friction pair will change the morphology of the photocatalyst, thereby generating more active sites. The $MoS_2@Ti_3C_2$ nanocomposite with a friction time of 60 min exhibited the best photocatalytic performance, and the removal rates of liquid paraffin and methyl orange were 80.5% and 98%, respectively. The improvement of the photocatalytic performance of the $MoS_2@Ti_3C_2$ nano-hybrid materials mainly can be ascribed to the following reason. First, MoS_2 grows vertically on Ti_3C_2 ensuring that more active sites can be exposed. Second, the formation of the heterojunction and Schottky junction reduce recombination of photogenerated electrons and holes. Third, the combination of the narrow band gap of MoS_2 and the wide band gap of TiO_2 increased the range of the solar absorption. In practical application, the $MoS_2@Ti_3C_2$ photocatalyst can directly decompose the waste oil into nonpolluting products under visible light irradiation. In this way, the waste oil can be directly discarded in nature without worrying about pollution (Jiao and Liu 2019).

Another common method for constructing binary MXene-based photocatalyst is to prepare MXene and semiconductor, respectively, and then assemble the MXene with semiconductor. In this way, the MXene-based heterojunction can be synthesized at a low temperature and the oxidation of MXene can be prevented. For example, the ultrasonic-assisted self-assembly method was used to prepare the Ti_3C_2/g-C_3N_4 nanocomposites (He et al. 2020). Due to the electrostatic attraction, there is a close contact interface between the Ti_3C_2 and g-C_3N_4 after combining the with Ti_3C_2 the g-C_3N_4. Such a compact interface can enhance the transfer rate of the charge carriers and reduce the recombination of the photogenerated electrons and holes. Under light irradiation, the 2D/2D Ti_3C_2/g-C_3N_4 heterostructure exhibited a higher diclofenac removal efficiency than that of the g-C_3N_4 photocatalyst. In addition, the Ti_3C_2/g-C_3N_4 system maintains a good photocatalytic performance in a wide pH range of 3–14. The enhanced photocatalytic performance can be ascribed to the 2D/2D interface between Ti_3C_2 and g-C_3N_4, which greatly promotes the migration and separation of photogenerated carriers in Ti_3C_2/g-C_3N_4. According to the quenching experiment and EPR results, the $^1O_2$ is the main reactant for diclofenac removal in the Ti_3C_2/g-C_3N_4 system, and three main ways to degrade diclofenac include the hydroxyl addition reaction, decarboxylation reaction, and dechlorination coupling reaction (Figure 8.4).

2D/2D Ti_3C_2/porous g-C_3N_4 composite was prepared by ultrasonic stripping and direct vacuum filtration (Liu et al. 2020). In the 2D/2D Ti_3C_2/porous g-C_3N_4 composite photocatalyst, the porous g-C_3N_4 nanolayer act as the visible light absorber, and the Ti_3C_2 nanolayer acts as the electron acceptor. The van der Waals heterojunction was formed between the Ti_3C_2 and porous g-C_3N_4, which facilitated the migration of photogenerated electrons from porous g-C_3N_4 to Ti_3C_2. With the 2D/2D Ti_3C_2/porous g-C_3N_4 composite as the photocatalyst, the degradation rate of phenol reaches as high as to 98%. Interestingly, with Ti_3C_2 as the electron storage material, can store excess photogenerated electrons can be store in Ti_3C_2 under light irradiation, and release them after exposure to electron acceptors in the dark. These released electrons can react with oxygen and water to produce $\cdot O_2^{2-}$ and $^{\bullet}OH$ radicals. Therefore, even in the absence of sunlight, 32% of phenol can be decomposed over 2D/2D Ti_3C_2/porous

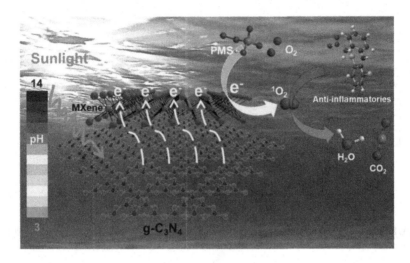

FIGURE 8.4 Mechanism of $Ti_3C_2/g\text{-}C_3N_4$ photocatalytic degradation of diclofenac (He et al. 2020).

$g\text{-}C_3N_4$ composite. In addition, the 2D/2D Ti_3C_2/porous $g\text{-}C_3N_4$ composite can also be used for degradation of various organic pollutants.

The thickness of a photocatalyst will affect the distance of photo-generated electron transferred to the surface of the material. Wang et al. (Wang, Shen, et al. 2020) prepared the $BiOCl/Ti_3C_2T_x$ composite through electrostatic self-assembly. The nano-thick BiOCl can be synthesized by controlling the solvothermal conditions. And the negative charged few-layered $Ti_3C_2T_x$ can be obtained by exfoliating the multilayer $Ti_3C_2T_x$, which was used to combine with BiOCl through the electrostatic self-assembly. The superior contact interface between the BiOCl and $Ti_3C_2T_x$ reduces the distance of electron transfer and ensures the rapid transfer of electrons from BiOCl to $Ti_3C_2T_x$. Therefore, the $BiOCl/Ti_3C_2T_x$ composite exhibits a high-degradation efficiency for p-nitrophenol. From the mechanism of $BiOCl/Ti_3C_2T_x$ for photocatalytic degradation of p-nitrophenol (Figure 8.5), the most active free radicals in this system are the hole-free radicals and superoxide free radicals.

MXene-based photocatalyst can also be synthesized by oxidizing the MXene itself. The metal elements on the MXene surface can be oxidized easily. Taking $Ti_3C_2T_x$ as an example, the Ti_3C_2/TiO_2 heterostructure can be obtained by exposing the $Ti_3C_2T_x$ to the $\cdot O^{2-}$ containing environment or calcining $Ti_3C_2T_x$ with an oxidant at high temperature. Chen prepared the $TiO_2/Ti_3C_2T_x$ composite by solvothermal method (Chen et al. 2020). In this way, the morphology and photocatalytic activity depend on the synthesis temperature (160–220 °C). The exposed (101) and (001) faces of in situ grown TiO_2 can transfer photogenerated electron-hole pairs and reduce their recombination by combining with the 2D MXene.

As shown in Figure 8.6, $TiO_2/Ti_3C_2T_x$ composites was synthesized by solvothermal method. In this $TiO_2/Ti_3C_2T_x$ photocatalytic system, the photogenerated electrons can transfer from the conduction band of TiO_2 to the layered $Ti_3C_2T_x$ under ultraviolet-light irradiation, which reduces the recombination of photogenerated

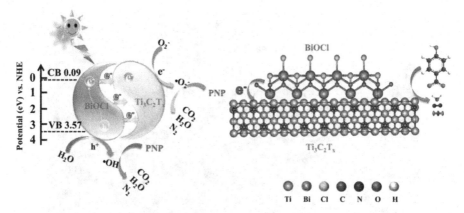

FIGURE 8.5 Photodegradation mechanism of BiOCl/$Ti_3C_2T_x$ (Wang, Shen, et al. 2020).

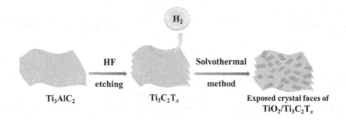

FIGURE 8.6 The schematic diagram for fabricating the facet exposed TiO_2/$Ti_3C_2T_x$ composites (Chen et al. 2020).

electron-hole pairs. Therefore, the photocatalytic degradation efficiency of TiO_2/$Ti_3C_2T_x$ composite for MO reached as high as to 92%. Under light irradiation, the photogenerated holes and electrons excited from the TiO_2 and react with water molecules to generate oxide ions and hydroxide radicals. In addition, the photogenerated electrons generated on the TiO_2 can transfer to the $Ti_3C_2T_x$, reducing the rapid recombination of photogenerated electron-hole pairs, which greatly improved the photocatalytic performance for the degradation of organic pollutants.

In summary, due to the electronic conductivity and high work function of MXenes, MXene have been widely used to combine with the semiconductor to form the binary MXene-based photocatalysts. The compact interface between the MXene and semiconductor greatly improve the separation of the photogenerated electrons and holes. Moreover, the formation of the Schottky junction at the MXene/semiconductor interface can inhibit the backflow of the electron from the MXene to the semiconductor, which enhances the performance of the photocatalytic degradation of the organic pollutant.

8.3 MULTICOMPONENT MXENE-BASED PHOTOCATALYST

In addition to the binary MXene-based photocatalyst, multicomponent MXene-based photocatalyst also shows the application prospect for photodegradation. In the multicomponent MXene-based photocatalyst system, there are multiple paths for the transfer path of the photogenerated electrons and holes between the semiconductors and the MXenes. With MXene cocatalyst and the heterostructure of the semiconductors, the separation of the photogenerated electrons and holes can be significantly enhanced (Li et al. 2021). Generally, MXenes participate in the electron transfer process, and the holes transfer between semiconductors in the multicomponent system (Wang, Song, and Zhang 2020).

Xu et al. synthesized a series of BiOBr/TiO$_2$/Ti$_3$C$_2$T$_x$ composites by a one-step hydrothermal method (Xu et al. 2020). Due to the synergistic effect of BiOBr, TiO$_2$, and Ti$_3$C$_2$T$_x$, the performance of photocatalytic rhodamine B degradation was greatly enhanced. The performance and mechanism of photocatalytic degradation of RhB in BiOBr/TiO$_2$/Ti$_3$C$_2$T$_x$ was shown in Figure 8.7. In terms of energy-level positions, the band gap of TiO$_2$ (~3.2 eV) is wider than that of BiOBr (~2.7 eV), and the positions of the conductive band (CB) and valence band (VB) corresponding to TiO$_2$ are both more negative than that of BiOBr. Therefore, in the BiOBr/TiO$_2$/Ti$_3$C$_2$T$_x$ system, the photogenerated electrons can transfer from the CB of the TiO$_2$ to the CB of the BiOBr, and then transfer to the Ti$_3$C$_2$T$_x$. In addition, the remaining Ti$_3$C$_2$T$_x$ in the BiOBr/TiO$_2$/Ti$_3$C$_2$T$_x$ ternary hybrid system can be used as an effective electron trap due to its inherent electrical conductivity to further inhibit the recombination of photoinduced electron-hole pairs. In addition, after coupling with the Ti$_3$C$_2$T$_x$, the absorption edge of BiOBr/TiO$_2$/Ti$_3$C$_2$T$_x$ nanocomposites extends to the visible light region, and the light-absorption intensity is enhanced compared with that of BiOBr, which is beneficial to the utilization of the solar energy. Therefore, the BiOBr/TiO$_2$/Ti$_3$C$_2$T$_x$ exhibited efficient photocatalytic performance for the rhodamine B degradation.

Moreover, the photogenerated holes on the VB of the BiOBr can transfer to the VB of the TiO$_2$, which will restrain the recombination of the photogenerated electrons and holes. The holes accumulated in the VB of TiO$_2$ have an excellent ability to directly oxidize the contaminants adsorbed on the surface of the photocatalyst. At the same

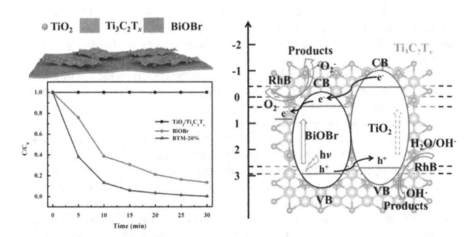

FIGURE 8.7 Performance and mechanism of rhodamine B degradation catalyzed by BiOBr/TiO$_2$/Ti$_3$C$_2$T$_x$ composite (Xu et al. 2020).

time, a large number of holes can also react with OH^- groups and H_2O molecules to generate $^{\bullet}OH$ radicals. Superoxide anions ($^{\bullet}O^{2-}$) may also be generated by the reaction of electrons in the CB of BiOBr with oxygen molecules adsorbed on the surface of the material or dissolved in water. Therefore, both $^{\bullet}OH$ radical and O^{2-} radical can be used as active substances to effectively degrade rhodamine B.

Two-dimensional $CdS@Ti_3C_2@TiO_2$ nanohybrid was synthesized by simple calcination and subsequent hydrothermal process (Liu et al. 2019). Compared with Ti_3C_2 and $Ti_3C_2@TiO_2$, the degradation of sulfachloropyrazine, phenol and several typical dyes were significantly enhanced by using the optimal $CdS@Ti_3C_2@TiO_2$ photocatalyst. In this ternary system, the Ti_3C_2 layer can act as a charge transport bridge between the $CdS-TiO_2$ Z-scheme system. Because the conduction band (CB) of TiO_2 (−0.25 V vs. NHE) is more negative than that of Ti_3C_2 (−0.15 V), photogenerated electrons were easily transferred to Ti_3C_2. In addition, the Schottky barrier formed at the Ti_3C_2/TiO_2 interface can prevent the electrons from flowing back into TiO_2. Moreover, the conduction band (CB) and valence band (VB) of CdS are located at −1.5 V and 0.8 V (vs. NHE), respectively. The photogenerated holes of CdS can migrate to the Ti_3C_2, leaving photoinduced electrons in the CB of CdS. Therefore, due to the formation of the Z-scheme heterostructure, the separation of electron-hole pairs is promoted, thereby improving the photocatalytic performance. In addition, because photogenerated electrons are accumulated in CB, which is more negative for CdS, and holes are located in VB, which is more positive for TiO_2, the $CdS@Ti_3C_2@TiO_2$ composite exhibited higher redox capacity opposed to pure materials.

Recently, it has been reported that the oxidized derivative of Ti_3C_2 unfolds better photocatalytic degradation performance of organic matter than the raw Ti_3C_2. For example, a Z-scheme photocatalyst of TiO_2/g-C_3N_4 immobilized by graphene derived from MXene was synthesized through a two-step calcination method (Wu et al. 2020). The graphene/TiO_2/g-C_3N_4 photocatalyst displayed enhanced photocatalytic activity for degradation of tetracycline, ciprofloxacin antibiotic, bisphenol A endocrine disruptor, and rhodamine B dye under visible light irradiation. As can be seen from Figure 8.8, the degradation process is mainly caused by $^{\bullet}O^{2-}$ and -OH, and these two active groups are produced on the VB of g-C_3N_4 and the CB of TiO_2, respectively. With the derived graphene as the electronic bridge, the photogenerated electrons on the TiO_2 can transfer to the graphene and combine with the holes from the g-C_3N_4, and the photoinduced electrons and holes was left on the CB of g-C_3N_4 and the VB of respectively, which will greatly enhance the photocatalytic performance for organic pollutants degradation.

Wang et al. reported a new quasi-core-shell In_2S_3/anatase TiO_2@metallic $Ti_3C_2T_x$ hybrid, which is composed of a Type II heterojunction and a non-noble metal-based Schottky junction. Due to the formation of Type II heterojunction and Schottky junction, the separation and diffusion of photogenerated electrons and holes was enhanced through multiple charge-transfer channels, which improved the photocatalytic degradation performance. The enhanced performance is due to the synergistic effect of the visible light absorption of In_2S_3, the upward-bending band of TiO_2, and the excellent conductivity of $Ti_3C_2T_x$. The separation of electrons and holes was conducive to the generation of oxygen radicals (such as O^{2-}) in the $Ti_3C_2T_x$ plane, thereby enhancing the photocatalytic degradation ability (Figure 8.9). It should be noted that

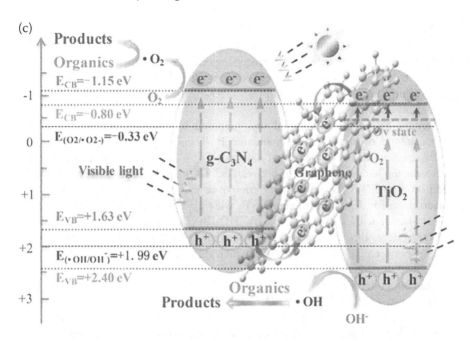

FIGURE 8.8 Schematic of photocatalytic process for organic pollutants degradation over Z-scheme heterojunction of Graphene/TiO$_2$/g-C$_3$N$_4$ under visible light irradiation (Wu et al. 2020).

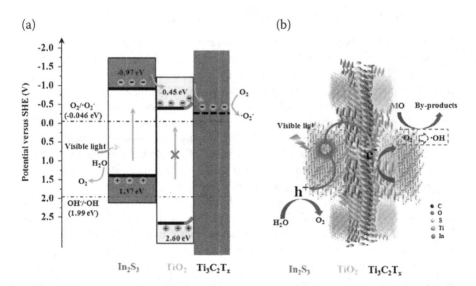

FIGURE 8.9 (a) Charge separation and transfer, and (b) proposed mechanism for pollutant degradation in the In$_2$S$_3$/TiO$_2$/Ti$_3$C$_2$T$_x$ system under visible light illumination (Wang et al. 2018).

the $In_2S_3/TiO_2/Ti_3C_2T_x$ hybrid photocatalyst exhibited enhanced photocatalytic performance, which is 3.2 and 6.2 times higher than that of In_2S_3 and $Ti_3C_2T_x$, respectively. Notably, the photocatalytic degradation performance of the optimal $In_2S_3/TiO_2/Ti_3C_2T_x$ hybrid has exceeded the In_2S_3/carbon nanotube (CNT), In_2S_3/reduced graphene oxide (rGO), In_2S_3/MoS_2, and In_2S_3/TiO_2 hybrids and many other In_2S_3-based photocatalysts (Wang et al. 2018).

Ternary $TiO_2/C/BiVO_4$ hybrid was prepared by Shi et al. and used for the photocatalytic degradation of Rhodamine B. The $TiO_2/C/BiVO_4$ composite was synthesized with two-dimensional layered Ti_3C_2 MXene as precursor, which overcame the lattice mismatch of between the TiO_2 and $BiVO_4$ (Shi et al. 2019). In addition, the $TiO_2/C/BiVO_4$ ternary composite exhibited enhanced photocatalytic performance for degrading rhodamine B, which is about four times and two times higher than that of $BiVO_4$ and binary $TiO_2/BiVO_4$ respectively. In the $TiO_2/C/BiVO_4$ system, the C layer derived from the Ti_3C_2 act as the electron transfer channel between the TiO_2 and $BiVO_4$, which significantly promoted the separation of photogenerated charge carriers and improved the photocatalytic performance for Rhodamine B degradation.

The construction of the ternary MXene-based heterostructure shows a wide range of application prospects in the field of photocatalytic degradation. In some cases, the ternary MXene-based photocatalyst exhibited superior photocatalytic degradation activity than that of binary heterostructure. However, the preparation process for the ternary MXene-based photocatalyst is more complicated and needs to be improved. In short, ternary MXene-based photocatalysts are worthy of deep exploration and will play an important role in the field of photocatalytic degradation in the near furture.

8.4 MODIFIED MXENE-BASED PHOTOCATALYSTS

The photocatalytic performance of the MXene-based photoctalysts can be further improved by doping with metal or nonmetal element (Schäfer et al. 2012; Dong et al. 2015). Due to its high ionization energy and electronegativity, MXene is easy to form covalent bonds with other compounds by obtaining electrons during the reaction. Generally speaking, element doping will improve the separation of electron-hole pairs, resulting in the enhanced photocatalytic activity of MXene-based photocatalysts. For example, element doping can reduce the band gap of semiconductors to extend their absorption range of visible light. In addition, when the extraneous element was doped in to the lattice of the semiconductor, an isolated impurity level can be formed in the valence band, which is beneficial for the separation of the photogenerated electron-hole pairs. Moreover, when the metal was loaded on the surface sites of MXene-based heterojunction, the electrons can flow from the MXene to metal due to their different Fermi energy level, which is also considered to be an effective means to further promote the separation of photo-generated charge carriers. Notably, the surface active adsorption and reaction sites of the MXene-based photocatalyst can also be modified by introducing organic functional group in to the heterostructure.

Ke et al. used isopropylamine (iPA) to modify Ti_3C_2 and prepared the N-doped Ti_3C_2, and then the $iN-Ti_3C_2/TiO_2$ composites was obtained through in situ oxidation of N-doped Ti_3C_2 (Ke et al. 2021). The iN-Ti_3C_2/TiO_2 hybrid photocatalyst displayed excellent performance for degradation of methylene blue under UV light irradiation, which is significantly higher than that of TiO_2, Ti_3C_2/TiO_2, dimethyl sulfoxide modified Ti_3C_2/TiO_2 hybrid, and hydrazine monohydrate modified Ti_3C_2/TiO_2 hybrid. In this iN-Ti_3C_2/TiO_2 system, N-doping improved the electrochemical performance of Ti_3C_2, promoted electron migration in Ti_3C_2, and inhibited the recombination of photogenerated electron-hole pairs. Under light irradiation, the electrons on the TiO_2 surface can react with the adsorbed oxygen to form O^{2-}, and O^{2-} can further react with H^+ and electrons to produce the active $^{\bullet}OH$. In addition, the photogenerated holes can also react with the adsorbed H_2O/OH^- on TiO_2 to generate $^{\bullet}OH$. The amount of $^{\bullet}OH$ dominate the photocatalytic performance of the methylene blue degradation.

In addition to the single-element doping, the synergistic effect of dual-element doping can greatly improve the photocatalytic performance of the MXene-based photocatalyst, in some cases. For example, Fang et al. (Fang et al. 2020) synthesized a new type of Yb^{3+}/Tm^{3+} co-doped Ti_3C_2/Ag/Ag_3VO_4 composite by in situ partial-reduction method and hydrothermal method. With the optimal Yb/Tm ratio (Y/T: 0.3/0.005), the Ti_3C_2/Ag/Ag_3VO_4 composite showed enhanced photocatalytic performance for the degradation of tetracycline and organic dyes under visible-light irradiation. When Ti_3C_2/Ag/Ag_3VO_4 composite was used as the photocatalyst, the removal rate of tetracycline and methylene blue in 20 minutes reached 97% and 99%, respectively, and the removal rate of rhodamine B in 25 minutes reached 96%, which was higher than that of Ag/Ag_3VO_4. Notably, the single doping of Yb^{3+} or Tm^{3+} ions has little effect on the final photocatalytic performance of the 4-Ti/Ag/Ag_3VO_4 composite. Interestingly, after simultaneously introducing Yb^{3+}/Tm^{3+}, the photocatalytic activity of the 4-Ti/Ag/Ag_3VO_4 composite is significantly improved. This excellent photodegradation performance is due to the enhanced light absorption and rapid charge-carrier separation, which was ascribed to the formation of intermediate energy state after co-doping Yb^{3+}/Tm^{3+} ions. However, the high content of Yb^{3+}/Tm^{3+} in the 4-Ti/Ag/Ag_3VO_4 composite will cause serious deterioration of photodegradation performance.

During the process of photocatalytic degradation, the photogenerated electrons were easily excited from the valence band (VB) of Ag_3VO_4 to the conduction band (CB), and then the holes were left in the VB. The holes produced in the VB of Ag_3VO_4 have strong oxidation property and can mineralize various pollutants in the water. Since the CB potential of Ag/Ag_3VO_4 (-0.53 eV vs. NHE) is greater than the reduction potential of O_2/$^{\bullet}O_2$ (–0.046 eV vs. NHE), the electrons accumulated on the CB of Ag_3VO_4 may be captured by the adsorbed O_2, resulting in the production of $^{\bullet}O^{2-}$ species. In addition, the electrons can also be immediately captured by the formed intermediate energy state, resulting in high charge-carrier separation efficiency. At the same time, it should be noted that photoinduced electrons can splash on the surface of Ti_3C_2 MXene and eventually be captured by oxygen molecules to form $\bullet O^2$. In summary, the synergistic effect of co-doping Yb^{3+}/Tm^{3+} can greatly improve the photocatalytic performance of the Ti_3C_2/Ag/Ag_3VO_4.

The C/TiO_2 derived from MXene can also exhibit excellent photocatalytic degradation performance after element doping. Huang et al. (Huang et al. 2019) used

Ti_3C_2 MXene as a carbon skeleton and a homogenous titanium source to obtain a two-dimensional layered nitrogen-doped carbon-supported titanium dioxide (N-TiO_2@C) through a one-step in-situ preparation method. Based on the negatively charged and easily oxidized characteristics of Ti_3C_2 MXene, it is assembled with nitrogen-containing cationic compounds through electrostatic interaction, and converted into nitrogen-containing carbon-supported TiO_2 in situ under the CO_2 atmosphere at 550 °C. For the N-TiO_2@C composite, lattice defects were caused by nitrogen doping, and the N 2p orbital overlaps the O 2p orbital, which causes the valence band to rise, and the bandgap of TiO_2 is expanded to the visible region.

Under visible-light irradiation, photogenerated electrons and holes were produced on N-doped TiO_2, and the photogenerated electrons quickly transferred to the N-doped carbon. In addition, the conduction band of TiO_2 (−0.3 eV) is higher than that of carbon (−0.08 eV), thus the carbon, as an excellent conductor, can quickly capture photogenerated electrons and increase the separation of the photoinduced electron-hole pairs. Subsequently, the O_2 was reduced by the photoinduced electrons to generate superoxide radicals (E (O_2/radical $^\bullet O^{2-}$) = −0.046 V, vs. NHE), and the OH^- was oxidized by the photoinduced holes to generate the $^\bullet OH$ radicals (E(radicals)$^\bullet OH/OH^-$) = 1.99 eV, vs. NHE). Finally, the hydroxyl radicals and superoxide radicals as used for the phenol degradation.

A series MXene-based photocatalyst by modifing Ti_2C MXene with TiO_2, Ag_2O, Ag, PdO, Pd, and Au (Wojciechowski et al. 2019) were synthesized. In the photocatalytic degradation process over the modified Ti_2C (Figure 8.10), the photogenerated electrons transferred from the conduction band of TiO_2 to the Ti_2C, extending the life of electrons and holes. Interestingly, the noble metal nanoparticles modified MXene did not exhibit the expected higher photoactivity. In terms of the carrier transfer path, firstly, the photogenerated electrons on TiO_2 will quickly migrate to the surface active

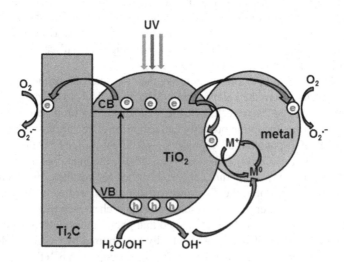

FIGURE 8.10 Proposed mechanism of the charge transfer in the Ti_2C MXene modified by metal oxides and metallic nanoparticles (metal = M = Ag, Pd, Au) (Wojciechowski et al. 2019).

sites of MXene and used for the production of •O2−. On the other hand, the H₂O was oxidized by the photogenerated holes to formed the •OH free radical, and the would migrate to the metal nanoparticles to oxidize the metal atoms into the M^+ cation, which would attract electrons and decrease the number of photogenerated electrons.

8.5 CONCLUSION AND PERSPECTIVES

This chapter summarized the application of different kinds of MXene-based photocatalysts in photocatalytic degradation. The MXene-based photocatalysts exhibited strong light-absorption ability and faster charge-carrier transfer rate, which can greatly enhance the separation of the photoctalytic electrons and holes. In addition, more reactive sites can be provided when the MXene-based photocatalysts were modified by element doping and surface engineering. Moreover, in some cases, the photocorrosion of the photocatalyst was prevented during the photocatalytic process when coupling with the MXene, which will enhance the stability of the photocatalyst and show broad prospects for photocatalytic degradation. However, the construction method, the stability, the photocatalytic performance of the MXene-based photocatalyst need to be further impoved to meet the requirement of the practical applications.

ACKNOWLEDGMENTS

This work was supported by the National Natural Science Foundation of China (22078074, 21938001), Guangxi Natural Science Foundation (2019GXNSFAA-245006, 2020GXNSFDA297007, 2016GXNSFFA380015), Special funding for 'Guangxi Bagui Scholars', and Scientific Research Foundation for High-level Personnel from Guangxi University.

REFERENCES

Akhade, Sneha A., Nirala Singh, Oliver Y. Gutiérrez, et al. 2020. Electrocatalytic hydrogenation of biomass-derived organics: A review. *Chemical Reviews* 120 (20):11370–11419.
Cai, Tao, Longlu Wang, Yutang Liu, et al. 2018. Ag₃PO₄/Ti₃C₂ MXene interface materials as a Schottky catalyst with enhanced photocatalytic activities and anti-photocorrosion performance. *Applied Catalysis B: Environmental* 239:545–554.
Cao, Yang, Yu Fang, Xianyu Lei, et al. 2020. Fabrication of novel CuFe₂O₄/MXene hierarchical heterostructures for enhanced photocatalytic degradation of sulfonamides under visible light. *Journal of Hazardous Materials* 387:122021.
Chen, Jin, Huiqi Zheng, Yang Zhao, et al. 2020. Preparation of facet exposed TiO₂/Ti₃C₂Tₓ composites with enhanced photocatalytic activity. *Journal of Physics and Chemistry of Solids* 145:109565.
Cheng, Lei, Xin Li, Huaiwu Zhang, and Quanjun Xiang. 2019. Two-dimensional transition metal MXene-based photocatalysts for solar fuel generation. *The Journal of Physical Chemistry Letters* 10 (12):3488–3494.
Dong, Yitong, Julius Choi, Hae-Kwon Jeong, and Dong Hee Son. 2015. Hot electrons generated from doped quantum dots via upconversion of excitons to hot charge carriers

for enhanced photocatalysis. *Journal of the American Chemical Society* 137 (16):5549–5554.

Fang, Hongjun, Yusong Pan, Haixian Yan, et al. 2020. Facile preparation of Yb^{3+}/Tm^{3+} co-doped $Ti_3C_2/Ag/Ag_3VO_4$ composite with an efficient charge separation for boosting visible-light photocatalytic activity. *Applied Surface Science* 527:146909.

He, Jie, Jingling Yang, Fengxing Jiang, Peng Liu, and Mingshan Zhu. 2020. Photo-assisted peroxymonosulfate activation via 2D/2D heterostructure of Ti_3C_2/g-C_3N_4 for degradation of diclofenac. *Chemosphere* 258:127339.

Hisatomi, Takashi and Kazunari Domen. 2019. Reaction systems for solar hydrogen production via water splitting with particulate semiconductor photocatalysts. *Nature Catalysis* 2 (5):387–399.

Hu, Anjun, Weiqiang Lv, Tianyu Lei, et al. 2020. Heterostructured $NiS_2/ZnIn_2S_4$ realizing toroid-like Li_2O_2 deposition in lithium–oxygen batteries with low-donor-number solvents. *ACS Nano* 14 (3):3490–3499.

Huang, Haowei, Bapi Pradhan, Johan Hofkens, Maarten B. J. Roeffaers, and Julian A. Steele. 2020. Solar-Driven metal halide perovskite photocatalysis: Design, stability, and performance. *ACS Energy Letters* 5 (4):1107–1123.

Huang, Huoshuai, Yun Song, Najun Li, et al. 2019. One-step in-situ preparation of N-doped $TiO_2@C$ derived from Ti_3C_2 MXene for enhanced visible-light driven photodegradation. *Applied Catalysis B: Environmental* 251:154–161.

Huang, Kelei, Chunhu Li, Haozhi Li, et al. 2020. Photocatalytic applications of two-dimensional Ti_3C_2 MXenes: A review. *ACS Applied Nano Materials* 3 (10):9581–9603.

Ji, Xiaozhong, Anthony Zuppero, Jawahar M. Gidwani, and Gabor A. Somorjai. 2005. Electron flow generated by gas phase exothermic catalytic reactions using a platinum–gallium nitride nanodiode. *Journal of the American Chemical Society* 127 (16):5792–5793.

Jiao, Songlong and Lei Liu. 2019. Friction-induced enhancements for photocatalytic degradation of $MoS_2@Ti_3C_2$ nanohybrid. *Industrial & Engineering Chemistry Research* 58 (39):18141–18148.

Ke, Tao, Shuyi Shen, Krishnamoorthy Rajavel, Kun Yang, and Daohui Lin. 2021. In situ growth of TiO_2 nanoparticles on nitrogen-doped Ti_3C_2 with isopropyl amine toward enhanced photocatalytic activity. *Journal of Hazardous Materials* 402:124066.

Khan, Muhammad Shuaib, Fengkai Zhang, Minoru Osada, Samuel S. Mao, and Shaohua Shen. 2020. Graphitic carbon nitride-based low-dimensional heterostructures for photocatalytic applications. *Solar RRL* 4 (8):1900435.

Khazaei, Mohammad, Avanish Mishra, Natarajan S. Venkataramanan, Abhishek K. Singh, and Seiji Yunoki. 2019. Recent advances in MXenes: From fundamentals to applications. *Current Opinion in Solid State and Materials Science* 23 (3):164–178.

Kong, Xianglong, Peng Gao, Rui Jiang, et al. 2020. Orderly layer-by-layered TiO_2/carbon superstructures based on MXene's defect engineeringfor efficient hydrogen evolution. *Applied Catalysis A: General* 590:117341.

Kumar, Ashish, Ajay Kumar, and Venkata Krishnan. 2020. Perovskite Oxide based materials for energy and environment-oriented photocatalysis. *ACS Catalysis* 10 (17):10253–10315.

Li, Bisheng, Shiyu Liu, Cui Lai, et al. 2020. Unravelling the interfacial charge migration pathway at atomic level in 2D/2D interfacial Schottky heterojunction for visible-light-driven molecular oxygen activation. *Applied Catalysis B: Environmental* 266:118650.

Li, Jiebo, Ruzhan Qin, Li Yan, et al. 2019. Plasmonic light illumination creates a channel to achieve fast degradation of $Ti_3C_2T_x$ nanosheets. *Inorganic Chemistry* 58 (11):7285–7294.

Li, Jing-Yu, Yue-Hua Li, Fan Zhang, Zi-Rong Tang, and Yi-Jun Xu. 2020. Visible-light-driven integrated organic synthesis and hydrogen evolution over 1D/2D CdS-$Ti_3C_2T_x$ MXene composites. *Applied Catalysis B: Environmental* 269:118783.

Li, Kaining, Sushu Zhang, Yuhan Li, Jiajie Fan, and Kangle Lv. 2021. MXenes as noble-metal-alternative co-catalysts in photocatalysis. *Chinese Journal of Catalysis* 42 (1):3–14.

Liu, Ning, Na Lu, HongTao Yu, Shuo Chen, and Xie Quan. 2020. Efficient day-night photocatalysis performance of 2D/2D Ti_3C_2/Porous g-C_3N_4 nanolayers composite and its application in the degradation of organic pollutants. *Chemosphere* 246:125760.

Liu, Qiaoran, Xiaoyao Tan, Shaobin Wang, et al. 2019. MXene as a non-metal charge mediator in 2D layered CdS@Ti_3C_2@TiO_2 composites with superior Z-scheme visible light-driven photocatalytic activity. *Environmental Science: Nano* 6 (10):3158–3169.

Liu, Yuanyue, Hai Xiao, and William A. Goddard. 2016. Schottky-barrier-free contacts with two-dimensional semiconductors by surface-engineered MXenes. *Journal of the American Chemical Society* 138 (49):15853–15856.

Naguib, Michael, Murat Kurtoglu, Volker Presser, et al. 2011. Two-dimensional nanocrystals produced by exfoliation of Ti_3AlC_2. *Advanced Materials* 23 (37):4248–4253.

Nguyen, Thang Phan, Dinh Minh Tuan Nguyen, Dai Lam Tran, et al. 2020. MXenes: Applications in electrocatalytic, photocatalytic hydrogen evolution reaction and CO_2 reduction. *Molecular Catalysis* 486:110850.

Peng, Jiahe, Xingzhu Chen, Wee-Jun Ong, Xiujian Zhao, and Neng Li. 2019. Surface and heterointerface engineering of 2D MXenes and their nanocomposites: Insights into electro- and photocatalysis. *Chem* 5 (1):18–50.

Prasad, Cheera, Xiaofei Yang, Qinqin Liu, et al. 2020. Recent advances in MXenes supported semiconductors based photocatalysts: Properties, synthesis and photocatalytic applications. *Journal of Industrial and Engineering Chemistry* 85:1–33.

Qin, Jiangzhou, Xia Hu, Xinyong Li, Zhifan Yin, Baojun Liu, and Kwok-ho Lam. 2019. 0D/2D AgInS2/MXene Z-scheme heterojunction nanosheets for improved ammonia photosynthesis of N_2. *Nano Energy* 61:27–35.

Ran, Jingrun, Guoping Gao, Fa-Tang Li, Tian-Yi Ma, Aijun Du, and Shi-Zhang Qiao. 2017. Ti_3C_2 MXene co-catalyst on metal sulfide photo-absorbers for enhanced visible-light photocatalytic hydrogen production. *Nature Communications* 8 (1):13907.

Rasool, Kashif, Ravi P. Pandey, P. Abdul Rasheed, Samantha Buczek, Yury Gogotsi, and Khaled A. Mahmoud. 2019. Water treatment and environmental remediation applications of two-dimensional metal carbides (MXenes). *Materials Today* 30:80–102.

Rojas, Sara and Patricia Horcajada. 2020. Metal–organic frameworks for the removal of emerging organic contaminants in water. *Chemical Reviews* 120 (16):8378–8415.

Ronchi, Rodrigo Mantovani, Jeverson Teodoro Arantes, and Sydney Ferreira Santos. 2019. Synthesis, structure, properties and applications of MXenes: Current status and perspectives. *Ceramics International* 45 (15):18167–18188.

Schäfer, Susanne, Sonja A. Wyrzgol, Roberta Caterino, et al. 2012. Platinum nanoparticles on gallium nitride surfaces: Effect of semiconductor doping on nanoparticle reactivity. *Journal of the American Chemical Society* 134 (30):12528–12535.

Shen, Jiyou, Jun Shen, Wenjing Zhang, et al. 2019. Built-in electric field induced CeO2/Ti_3C_2-MXene Schottky-junction for coupled photocatalytic tetracycline degradation and CO_2 reduction. *Ceramics International* 45 (18, Part A):24146–24153.

Shi, L., C. L. Xu, D. X. Jiang, et al. 2019. Enhanced interaction in TiO2/BiVO4 heterostructures via MXene Ti_3C_2-derived 2D-carbon for highly efficient visible-light photocatalysis. *Nanotechnology* 30 (7):11.

Sinopoli, Alessandro, Zakarya Othman, Kashif Rasool, and Khaled A. Mahmoud. 2019. Electrocatalytic/photocatalytic properties and aqueous media applications of 2D transition metal carbides (MXenes). *Current Opinion in Solid State and Materials Science* 23 (5):100760.

Sun, Bin, Furong Tao, Zixuan Huang, et al. 2021. Ti_3C_2 MXene-bridged Ag/Ag3PO4 hybrids toward enhanced visible-light-driven photocatalytic activity. *Applied Surface Science* 535:147354.

Tang, Rongdi, Sheng Xiong, Daoxin Gong, et al. 2020. Ti_3C_2 2D MXene: Recent progress and perspectives in photocatalysis. *ACS Applied Materials & Interfaces* 12 (51):56663–56680.

Tang, Xiao, Dong Zhou, Peng Li, et al. 2020. MXene-based dendrite-free potassium metal batteries. *Advanced Materials* 32 (4):1906739.

Vitale, Vincenzo, Alessandro Curioni, and Wanda Andreoni. 2008. Metal–carbon nanotube contacts: The link between Schottky barrier and chemical bonding. *Journal of the American Chemical Society* 130 (18):5848–5849.

Wang, Congjun, Juan Shen, Rigui Chen, Fang Cao, and Bo Jin. 2020. Self-assembled BiOCl/$Ti_3C_2T_x$ composites with efficient photo-induced charge separation activity for photocatalytic degradation of p-nitrophenol. *Applied Surface Science* 519:146175.

Wang, Dali, Jingbei Ren, Zongyi Tan, and Jing You. 2020. Gut microbial profiles in nereis succinea and their contribution to the degradation of organic pollutants. *Environmental Science & Technology* 54 (10):6235–6243.

Wang, Hou, Yuanmiao Sun, Yan Wu, et al. 2019. Electrical promotion of spatially photo-induced charge separation via interfacial-built-in quasi-alloying effect in hierarchical $Zn_2In_2S_5/Ti_3C_2(O, OH)_x$ hybrids toward efficient photocatalytic hydrogen evolution and environmental remediation. *Applied Catalysis B: Environmental* 245:290–301.

Wang, Hou, Yan Wu, Tong Xiao, et al. 2018. Formation of quasi-core-shell In_2S_3/anatase TiO_2@metallic $Ti_3C_2T_x$ hybrids with favorable charge transfer channels for excellent visible-light-photocatalytic performance. *Applied Catalysis B: Environmental* 233: 213–225.

Wang, Runmei, Kevin C. Jones, and Hao Zhang. 2020. Monitoring organic pollutants in waters using the diffusive gradients in the thin films technique: Investigations on the effects of biofouling and degradation. *Environmental Science & Technology* 54 (13):7961–7969.

Wang, Xiao, Shuyan Song, and Hongjie Zhang. 2020. A redox interaction-engaged strategy for multicomponent nanomaterials. *Chemical Society Reviews* 49 (3):736–764.

Wojciechowski, Tomasz, Anita Rozmysłowska-Wojciechowska, Grzegorz Matyszczak, et al. 2019. Ti_2C MXene modified with ceramic oxide and noble metal nanoparticles: Synthesis, morphostructural properties, and high photocatalytic activity. *Inorganic Chemistry* 58 (11):7602–7614.

Wu, Zhibin, Yunshan Liang, Xingzhong Yuan, et al. 2020. MXene Ti_3C_2 derived Z–scheme photocatalyst of graphene layers anchored TiO_2/g–C_3N_4 for visible light photocatalytic degradation of refractory organic pollutants. *Chemical Engineering Journal* 394:124921.

Xiao, Xu, Hao Wang, Weizhai Bao, et al. 2019. Two-dimensional arrays of transition metal nitride nanocrystals. *Advanced Materials* 31 (33):1902393.

Xie, Xiuqiang and Nan Zhang. 2020. Positioning MXenes in the photocatalysis landscape: Competitiveness, challenges, and future perspectives. *Advanced Functional Materials* 30 (36):2002528.

Xu, Tianxiang, Jiapei Wang, Ye Cong, et al. 2020. Ternary BiOBr/TiO_2/$Ti_3C_2T_x$ MXene nanocomposites with heterojunction structure and improved photocatalysis performance. *Chinese Chemical Letters* 31 (4):1022–1025.

Yun, Taeyeong, Hyerim Kim, Aamir Iqbal, et al. 2020. Electromagnetic shielding of monolayer mxene assemblies. *Advanced Materials* 32 (9):1906769.

Zhang, Chuanfang, Babak Anasori, Andrés Seral-Ascaso, et al. 2017. Transparent, flexible, and conductive 2D titanium carbide (MXene) films with high volumetric capacitance. *Advanced Materials* 29 (36):1702678.

Zhang, Huiru, Yinhua Wan, Jianquan Luo, and Seth B. Darling. 2021. Drawing on membrane photocatalysis for fouling mitigation. *ACS Applied Materials & Interfaces* 13 (13): 14844–14865.

Zhang, Huoli, Man Li, Jianliang Cao, et al. 2018. 2D a-Fe$_2$O$_3$ doped Ti$_3$C$_2$ MXene composite with enhanced visible light photocatalytic activity for degradation of Rhodamine B. *Ceramics International* 44 (16):19958–19962.

Zhao, Di, Zewen Zhuang, Xing Cao, et al. 2020. Atomic site electrocatalysts for water splitting, oxygen reduction and selective oxidation. *Chemical Society Reviews* 49 (7):2215–2264.

Zhao, Meng-Qiang, Chang E. Ren, Zheng Ling, et al. 2015. Flexible MXene/carbon nanotube composite paper with high volumetric capacitance. *Advanced Materials* 27 (2):339–345.

Zhao, Meng-Qiang, Xiuqiang Xie, Chang E. Ren, et al. 2017. Hollow MXene spheres and 3D macroporous MXene frameworks for Na-Ion storage. *Advanced Materials* 29 (37): 1702410.

9 Application of MXene-Based Photocatalyst in Other Photocatalytic Fields

Chengzheng Men and Tongming Su
School of Chemistry and Chemical Engineering, Guangxi University

CONTENTS

9.1 INTRODUCTION

MXenes, as a new two-dimension material with excellent electronic conductivity (Kim and Alshareef 2020; Lim et al. 2020; Lipatov et al. 2021), abundant terminal hydrophilic groups (Yu et al. 2019), numerous active sites (Zuo et al. 2020), large specific surface area (Peng et al. 2019), excellent mechanical properties (Firestein et al. 2020), thermal conductivity, and other properties (Hasan, Hossain, and Chowdhury 2021), have been widely used for a variety of applications. In the field of photocatalysis, MXenes can combine with other semiconductors to accelerate the separation and transmission of the photogenerated charge carriers, thus improving the photocatalytic performance of the semiconductor photocatalysts. In addition to the application of MXenes as co-catalysts in the fields of photocatalytic water-splitting (Ran et al. 2017), CO_2 reduction (Zhao et al. 2020), and degradation of organic matter (Feng et al. 2021), MXenes has also been used in photocatalytic nitrogen fixation, H_2O_2 production, and sterilization in recent years. In this chapter, we focus on the recent applications of MXenes in photocatalytic nitrogen fixation, H_2O_2 preparation, and antimicrobial activities. The reaction mechanisms are

DOI: 10.1201/9781003156963-9

discussed in detail, and some bottlenecks and problems still existing in these aspects are reviewed and summarized. This chapter aims to broaden the multiple applications of MXenes in the field of photocatalysis.

MXenes has gained comprehensive attention in the field of photocatalytic antimicrobial. In modern medicine, antibiotics are widely used to treat bacterial infectious diseases, which can cause damage to the human immune system (Schwaber, De-Medina, and Carmeli 2004). Intriguingly, photocatalysis technology can use superoxide radicals produced during light excitation to kill bacteria, thus avoiding the harmful effects of oral drugs (Wang, Zhang, et al. 2019; Han et al. 2020; Xia et al. 2020). Due to the local-surface plasmonic resonance of MXene, which can improve the overall light-response range and light-absorption capacity, MXene can be used as a the co-catalyst of photocatalyst for photocatalytic antimicrobial. In addition, the electron harvesting capability and good electrical conductivity of MXene enable the enhancement of photocatalytic performance by accelerating the transfer rate of photoinduced charge carriers.

9.2 PHOTOCATALYTIC NITROGEN FIXATION

Natural nitrogen (N_2) occupies the highest proportion in the earth's atmosphere; however, it can only be used for biosynthesis and is hard to convert into NH_3 or NO_3^-. Notably, NH_3 and NO_3^- can be obtained from the photocatalytic nitrogen fixation (Kärkäs 2017; Medford and Hatzell 2017; Li, Wang, et al. 2021). At present, most of the NO_x discharged from industry can be oxidized to NO_3^-. Photocatalytic oxidation technology is an efficient method to convert the nitrogen and oxygen pollutants into chemical materials with solar energy as the driving force. In addition, photocatalytic reduction of N_2 to produce ammonia (NH_3) is also a viable way to make organic fertilizer from N_2 (Xiong et al. 2020). In industry, more than 90% of NH_3 was produced by the Haber-Bosch process, and those produced NH_3 can be used as a precursor to obtain nitrogen fertilizer. However, the high cost of the Haber-Bosch process is a problem in this route. In practice, due to the strong bond N≡N (with a bond energy of 945 kJ mol^{-1}) and large ionization of N_2 (Wang and Meyer 2019; Elishav et al. 2020), it is very difficult to convert N_2 directly to NH_3. At present, the electrocatalysis and photocatalysis technology for nitrogen fixation faces poor catalytic activity, and the production of NH_3 or NO_3^- is far from meeting the industrial requirements. In recent studies, heteroatom doping, vacancy engineering, formation of interface, and surface modification have been used to optimize the intrinsic activity of photocatalysts (Li 2018). Due to the requirement to the active site, precious metal catalysts have attracted more and more attention in the past few years, while the expensive cost greatly limits the large-scale commercial application of precious metals.

The industrial process for synthesis of ammonia, which uses the Haber-Bosch process, is carried out at high temperatures (673 K) and high pressures (200 atm), and a lot of pollutants are produced. Photocatalytic technology provides a novel clean route for photocatalytic-nitrogen fixation. For example, in recent years, TiO_2 (Xie et al. 2019), g-C_3N_4 (Li, Yu, et al. 2021), and ZnO (Xing et al. 2018), etc., have been developed into suitable photocatalysts for photocatalytic synthesis of ammonia.

However, due to the rapid recombination of electron-hole pairs and the poor adsorption capacity of nitrogen molecules on the photocatalyst surface, the activity of nitrogen fixation by photocatalytic technology is still very low.

As a novel type of transition metal carbide and nitride, MXene has become a good substitute of precious metal for improving photocatalytic nitrogen fixation performance due to its high specific surface area, adjustable elemental composition, specific functional groups, excellent conductivity, hydrophilicity, and suitable Fermi-level. The compact interface formed between the semiconductor and MXenes can accelerate the transfer of large amounts of electrons to the photocatalyst surface, and the surface of the 2D MXene can provide more active sites for reduction of N_2. Therefore, MXene is a highly efficient and promising co-catalyst for improving the photocatalytic activity of N_2 reduction.

At present, most of the applications of MXenes in photocatalytic nitrogen fixation serve as the co-catalysts, which play a good role in adsorption and inhibiting the recombination of photogenerated charge carriers. For example, Liao et al. reported the formation of binary Ti_3C_2 MXenes-P25 heterostructures by electrostatic assembly (Liao et al. 2020). Interestingly, 2D Ti_3C_2 MXene can significantly improve the photocatalytic activity of P25 for NH_3 synthesis. Interestingly, when using H_2O as a sacrificial agent, the NH_3 yield of 6% Ti_3C_2 MXenes-P25 was five times than that of P25. However, when CH_3OH was used as a proton source and hole scavenger, the NH_3 yield was only 1.4 times that of P25. This result showed that with H_2O as a proton source and a hole scavenger, Ti_3C_2 MXene can accelerate the charge-carrier separation and promote N_2 adsorption. According to the density functional theory calculation, N_2 molecules on Ti_3C_2 MXenes surface are more thermodynamically activated than those on pure TiO_2 and oxygen-rich vacancy TiO_2 surface. During the photocatalytic synthesis of NH_3, Ti_3C_2 MXene not only enhanced the separation of the photogenerated electron-hole pairs, but also promoted the chemisorption and activation of N_2. The photocatalytic mechanism of the Ti_3C_2 MXenes-P25 heterostructures can be seen in Figure 9.1.

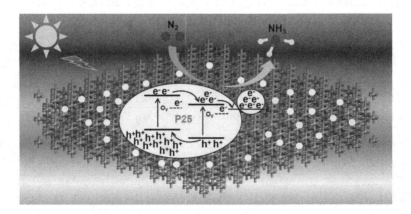

FIGURE 9.1 Schematic illustration for the promotion effect of 2D-layered Ti_3C_2 MXenes on P25 for photocatalytic NH_3 synthesis from N_2 (Liao et al. 2020).

In addition to the adsorption capacity of MXene and the ability to promote the separation of photogenerated carriers, the synergistic effect of MXene and other cocatalysts is also an effective way to promote the photocatalytic nitrogen-fixation activity. For example, Gao et al. explored the relationship between the adsorption of reactants (N_2) and products (NH_3) on the catalyst surface and the photocatalytic nitrogen-fixation activity (Gao et al. 2021). In this work, the $Ti_3C_2@TiO_2$ binary heterostructure was constructed by in situ growing of TiO_2 nanoparticles on the surface of MXene nanosheets first, and then, Co co-catalyst was also introduced into the system. The synergistic effect of the two co-catalysts regulated the surface reactivity of nitrogen and water. The in situ growing technology can make TiO_2 layer-by-layer coating on the surface of the Ti_3C_2 layer. The close contact between the TiO_2 and the Ti_3C_2 promoted the rapid transport and separation of charge carriers. In addition, the Co co-catalyst could adjust the adsorption equilibrium of N_2 and NH_3 on the catalyst surface, thus promoting the product desorption on the catalyst and improving the conversion efficiency of the active center. The yield of ammonia in pure water is significantly enhanced without the addition of any hole sacrifice agent due to the excellent conductivity, efficient carrier separation, and suitable nitrogen adsorption of MXene.

The mechanism of the MXene/TiO_2/Co for photocatalytic NH_3 synthesis from N_2 was described in Figure 9.2. Under the light irradiation, electrons in the valence band of TiO_2 can be activated to the conduction band, leaving holes in the valence band. Due to the formation of the Ti_3C_2/TiO_2 heterojunction, the electrons and holes can transfer to the surface effectively. Subsequently, the electrons were captured by the oxygen vacancy and the adsorbed nitrogen on the surface was reduced. In addition, the Co cocatalyst can regulate the binding of catalyst to reactants and products, resulting in the timely desorption of products on the active sites, which was beneficial to the regeneration of active sites.

In addition to oxide/MXene-based catalysts, sulfides with adjustable morphology and band gap combining with MXene have also shown excellent performance in photocatalytic nitrogen fixation. For example, 0D $AgInS_2$ nanoparticles can be produced on the surface of 2D Ti_3C_2 MXenes to form the 0D/2D $AgInS_2$/Ti_3C_2

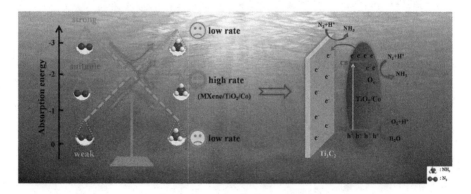

FIGURE 9.2 Schematic illustration for the promotion effect of MXene/TiO_2/Co for photocatalytic NH_3 synthesis from N_2 (Gao et al. 2021).

heterojunction by the in situ growth method (Qin et al. 2019). This 0D/2D AgInS$_2$/Ti$_3$C$_2$ heterostructure exhibited the advantages of short charge-transfer distance and large specific surface area, while 2D Ti$_3$C$_2$ nanosheets can provide more contact area and sites to capture N$_2$. In addition, Ti$_3$C$_2$ can also continuously accelerate the accumulation of photogenerated electrons on its surface, which can enhance the photocatalytic reduction of nitrogen dioxide under visible light irradiation.

Interestingly, the optical conductivity in the middle of the visible spectrum can be changed, adjusting the number and type of surface functions of Ti$_3$C$_2$. The absorption edge of MXene tends to be zero due to its metallic nature. In this case, after AgInS$_2$ was in close contact with Ti$_3$C$_2$, a Schottky junction is formed between AgInS$_2$ and Ti$_3$C$_2$, and Ti$_3$C$_2$ can be used as an effective electron trap to promote the separation of photogenerated charge carriers of AgInS$_2$. Notably, MXene may provide a feasible kinetic approach for the effective nitrogen fixation. As shown in Figure 9.3, the N≡N bond length has been strongly elongated from 1.098 Å to 1.334 Å, and the large adsorption energy ($E_{ad} = -5.20$ eV) indicated that N$_2$ activation can be spontaneously achieved when it is chemically adsorbed on MXene nanosheets.

In addition to the conversion of N$_2$ to NH$_3$, photocatalytic oxidation of oxynitride pollutants to NO$_3^-$ is also an efficient way for nitrogen fixation. For example, Wang et al. reported that Ti$_3$C$_2$ MXene acted as the adsorbent and photocatalytic

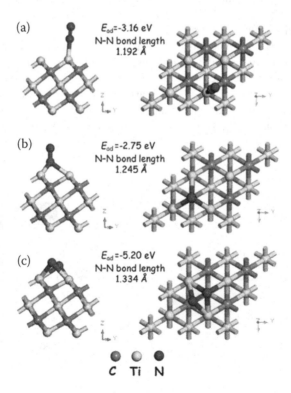

FIGURE 9.3 Different adsorption configurations of N$_2$ molecules with their corresponding N–N bond lengths in the Ti$_3$C$_2$ (001) surface with a ($2 \times 2 \times 1$) supercell (Qin et al. 2019).

oxidation center in the MIL-100(Fe)/Ti$_3$C$_2$ MXene composite, which displayed the efficient photocatalytic conversion of NO to NO$_3^-$. Therefore, MIL-100(Fe)/Ti$_3$C$_2$ MXene composite exhibited high thermal stability and significant nitrogen-fixation activity under visible light. The performance of the photocatalytic oxidation of NO was four times and three times higher than that of pure Ti$_3$C$_2$ and MIl-100(Fe) samples (Wang, Zhao, et al. 2019). The enhancement of NO photocatalytic oxidation is mainly due to the existence of Ti$_3$C$_2$ MXene. As shown in Figure 9.4, under visible light, the photogenerated electrons excited on MIL-100(Fe) can be transferred to the surface of Ti$_3$C$_2$ through the contact interface. This is mainly due to the fact that the Fermi level of Ti$_3$C$_2$ is much lower than that of MIL-100(Fe), and the formation of Schottky contact between the Ti$_3$C$_2$ and MIl-100(Fe) provided an effective channel for the rapid charge carrier transfer in the photocatalytic reaction process, thus enhancing the separation of the photogenerated charge carriers. In addition to electron harvesting, Ti$_3$C$_2$ also enhanced the light absorption of the photocatalyst and adsorption capacity of the NO, which contributes to the conversion of oxynitride pollutants to nitrate.

In general, MXene in a smaller size can expose more active sites, which is conducive to adsorption of reactants and the photocatalytic reaction. Therefore, MXene quantum dots have attracted great attention for photocatalytic conversion of NO to NO$_3^-$ (Wang et al. 2020). For example, Ti$_3$C$_2$ QDs/SiC composite by coupling of SiC and Ti$_3$C$_2$ MXene quantum dots (QDs) is achieved through the electrostatic self-assembly method. The efficiency of photocatalytic NO oxidation over the Ti$_3$C$_2$ QDs/SiC heterojunction reached 74.6% under visible light irradiation, which was 3.1 times and 3.7 times higher than that of the pure Ti$_3$C$_2$ MXene quantum dots and SiC, respectively. Moreover, Ti$_3$C$_2$ QDs/SiC composite not only has a narrower band gap than SiC, but it also has a stronger oxidizing capacity of VB holes. In this system, the factors that affect the overall activity of photocatalytic nitrogen fixation are S$_{BET}$, surface groups, and ·O$_2^-$. The surface area of Ti$_3$C$_2$

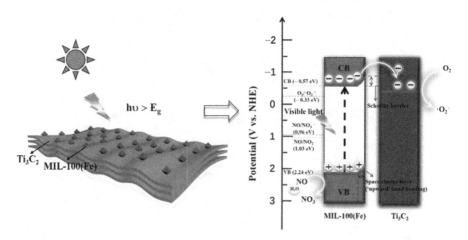

FIGURE 9.4 Proposed mechanism for photocatalytic oxidation of NO by the Ti$_3$C$_2$/MIL-100(Fe) Schottky catalyst under visible light illumination (Wang, Zhao, et al. 2019).

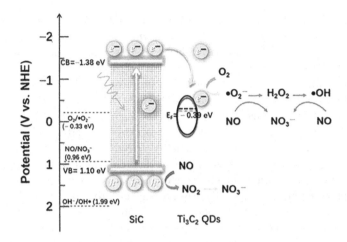

FIGURE 9.5 Photocatalytic mechanism of NO degradation by Ti_3C_2 QDs/SiC heterojunction under visible light irradiation (Wang et al. 2020).

reaches 362.31 m^2 g^{-1}, and the specific surface area of the Ti_3C_2 QD$_S$/SiC composite was increased by two times compared with that of pure SiC. Proper pore size is also conducive to the contact between pollutants and photocatalyst. In addition, photoelectrons can also reduce O_2 to promote the production of O_2^-, which in turn can participate in the NO conversion reaction. As shown in Figure 9.5, an efficient electron transfer channel was formed between the Ti_3C_2 quantum dots and SiC, which greatly enhanced the performance of the photocatalytic NO conversion.

In summary, photocatalytic nitrogen-fixation technology can solve both energy consumption and environmental pollution problems in industry (Zhang, Ward, and Sigman 2020). When MXene was used as co-catalyst in the photocatalytic nitrogen fixation, the transfer of photogenerated charge carriers was accelerated, and the more active sites were created, which can improve the performance of the photocatalytic nitrogen fixation. Despite the fast development of the MXene-based photocatalyst for the nitrogen fixation, there are still challenges for the industrial application of this technology. For example, the influence of the surface group of MXene on the adsorption capacity of nitrogen sources has not been fully understood, that is, which groups can promote the adsorption of nitrogen sources (NO, N_2, etc.) and which groups are not conducive to the adsorption of nitrogen sources is not clear yet. Furthermore, the construction method of the MXene-based photocatalysts needs to be optimized for the photocatalytic nitrogen-fixation reaction. Therefore, more efforts are needed to promote the development of the Mxene-based nitrogen-fixing photocatalysts.

9.3 PHOTOCATALYTIC PRODUCTION OF H_2O_2

As is well-known, H_2O_2 is a clean, environmentally friendly, and multifunctional oxidant, which produces H_2O after the reaction. In industrial production and daily life, H_2O_2 has been widely used in bleaching, disinfection, wastewater treatment, and fuel cell (Sun, Han, and Strasser 2020; Zhou et al. 2021). However, the main industrial synthesis method of H_2O_2 is high energy consumption, and toxic by-products are produced simultaneously; therefore, it is necessary to find a green and efficient synthesis path for the production of H_2O_2 (Wang et al. 2021). Photocatalysis technology is considered as a promising method for the production of H_2O_2 because the reduction potential of O_2/H_2O_2 (0.695 V vs. SHE) (Moon et al. 2017) is lower than the valence-band potential of many semiconductors. In addition, photocatalytic reduction of O_2 to H_2O_2 is a process of converting solar energy into chemical energy, which is environmentally friendly. Moreover, photocatalytic technology is a low-cost method for H_2O_2 production, which is expected to replace the commonly used anthraquinone method in industry. In the process of photo-catalytic preparation of H_2O_2, the photogenerated electrons in the conduction band generated by the semiconductor photocatalyst can reduce O_2 to produce H_2O_2. However, in this process, the low separation and transfer efficiency of photo-generated carriers is still the main reason for the low activity of the photocatalyst.

By virtue of its superior conductivity, numerous surface hydrophilic groups, abundant active sites and electron capture ability, MXene can greatly promote the separation of photogenerated carriers, thus significantly improving the photo-catalytic activity for the production of H_2O_2. In the process of photoreduction production of H_2O_2, MXene can effectively inhibit the recombination of photo-generated electron-hole pairs, accelerate the electron-transfer process, improve the performance of the catalyst, and improve the reaction rate. In addition, the abundant surface groups on MXene can promote the adsorption of reactants. Furthermore, surface groups on MXene are regarded as the active sites for the H_2O_2 production. Some MXene-based photocatalysts were discovered for the photocatalytic pro-duction of H_2O_2 reaction and proved that MXene is an excellent co-catalyst.

Constructing a built-in electric field at the interface of the MXene-based pho-tocatalyst composites is an effective way to promote the migration of photo-generated carriers. Yang et al. reported Ti_3C_2/porous g-C_3N_4 composite nanosheets constructed by the electrostatic assembly method. The Ti_3C_2/porous g-C_3N_4 com-posite exhibited good photocatalytic activity for the production of H_2O_2 (Yang et al. 2019). Interestingly, the introduction of Ti_3C_2 can not only promote the photo-generated charge carrier separation on g-C_3N_4 under visible light, but also promote the transformation of reactants and the production of H_2O_2. In the process of photocatalytic reaction, the space charge carriers transfer between Ti_3C_2/g-C_3N_4 interface and the construction of Schottky barrier is favorable for the separation of photogenerated carriers.

As shown in Figure 9.6, the ultraviolet photoelectron spectroscopy (UPS) de-monstrated that the cut-off edge energies of Ti_3C_2, porous g-C_3N_4 and 2 wt% Ti_3C_2/g-C_3N_4 are 17.89, 18.67, and 18.46 eV, respectively (Yang et al. 2019). Thus, according to Einstein's law of photoelectric emission, the surface work functions of these three samples can be calculated to be 3.31, 2.53, and 2.74 eV, respectively. Since the work function of Ti_3C_2 is greater than that of g-C_3N_4, the electrons of g-C_3N_4 will transfer to Ti_3C_2 after the Ti_3C_2/g-C_3N_4 heterostructure was formed, and the energy band of g-C_3N_4 bend upward in this process until the Fermi level of the Ti_3C_2 and g-C_3N_4 reach equilibrium, finally forming a Schottky barrier at the

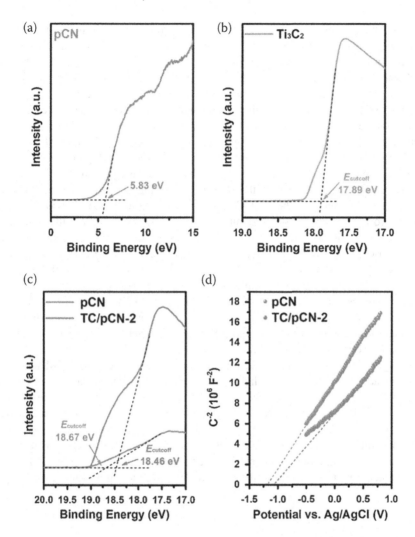

FIGURE 9.6 UPS spectra of (a) porous g-C$_3$N$_4$ and (b) Ti$_3$C$_2$; (c) UPS spectra and (d) Mott-Schottky plots of porous g-C$_3$N$_4$ and 2 wt% Ti$_3$C$_2$/g-C$_3$N$_4$ (Yang et al. 2019).

heterostructure interface. Due to the existence of the Schottky barrier, the electrons captured by Ti$_3$C$_2$ cannot be reflux into the conduction band of porous g-C$_3$N$_4$ and was used for the photocatalytic reaction. In addition, the slope of Mott-Schottky curve indicated that the porous g-C$_3$N$_4$ is an n-type semiconductor (Figure 9.6d). In short, the Ti$_3$C$_2$ plays a positive role in enhancing the photogenerated charge-carrier separation in g-C$_3$N$_4$, which is favorable for H$_2$O to obtain photogenerated electrons on the surface of Ti$_3$C$_2$ to produce H$_2$O$_2$.

Chen et al. (Chen et al. 2021) prepared Ti$_3$C$_2$/TiO$_2$ photocatalyst by simple impregnation method, and the H$_2$O$_2$ generation rate (179.7 μmol L^{-1} h^{-1}) of the Ti$_3$C$_2$/TiO$_2$ composite with the optimal Ti$_3$C$_2$ content exhibited more than 21 times higher than that of P25. In the reaction process, the existence of superoxide radical indicated that the generation of H_2O_2 is actually a two-step process; that is, the oxygen was converted to superoxide radical and then to hydrogen peroxide, and these two conversion processes occur on the surface of Ti_3C_2. It is worth mentioning that the production and decomposition of H_2O_2 usually occur simultaneously during the photocatalytic-reaction process. In the presence of P25, H_2O_2 can be completely decomposed in a relatively short time; this is mainly due to the Ti-OH formed by P25, which immediately reacts with H_2O_2 to form the hydrogen peroxide complex and reduces the production of H_2O_2. However, after loading with MXene, the decomposition rate of H_2O_2 under UV light is significantly reduced. Figure 9.7 shows the production process of H_2O_2 on the Ti_3C_2/TiO_2 heterostructure. In the Ti_3C_2/TiO_2 system, Ti_3C_2 greatly promotes the transfer of photogenerated electrons, thus inhibiting the formation of the Ti–OOH bond. In addition, the Ti_3C_2 with a higher surface-work function can serve as an electron container to promote the separation and transfer of photogenerated electrons and holes. Furthermore, the specific surface area of Ti_3C_2/TiO_2 composite is larger than that of P25, which provides more active sites for H_2O_2 production. Ti_3C_2 can also inhibit the decomposition of H_2O_2 and promotes the two-step O_2 reduction process to generate H_2O_2.

The synergistic effect of the MXene with other co-catalysts was also studied for photocatalytic H_2O_2 production. For example, the C_3N_4/Ti_3C_2 heterojunction was prepared and decorated with per-6-thio-β-Cyclodextrin (SH-β-CD)-protected platinum nanocluster (Pt@β-CD NCs), which exhibited excellent photocatalytic performance for H_2O_2 production (Zhu et al. 2021). In this system, the hybridization of MXene with C_3N_4 greatly improved visible light absorption and spatial separation of photogenerated charge carriers. Furthermore, Pt@β-CD can not only accelerate the electron transfer and provide active sites for H_2O_2 production, but it also provide a rich "delivery channel" to promote reactant (O_2) diffusion. In the process of photocatalytic O_2 reduction to H_2O_2, there are two main reaction ways, namely, two-electron reduction and continuous single-electron reduction of O_2. In the Pt@β-CD/C_3N_4 heterosystem, the introduction of the O_2^- scavenger resulted in a significant decrease

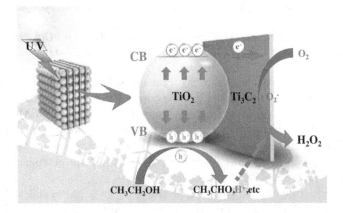

FIGURE 9.7   The possible mechanism of photocatalytic H_2O_2 production from Ti_3C_2/TiO_2 photocatalyst (Chen et al. 2021).

in H_2O_2 production, indicating that the single electron reduction of O_2 is the main way for photocatalytic production of H_2O_2 over Pt@β-CD/C_3N_4-Ti_3C_2 photocatalyst.

The reaction path from O_2 to H_2O_2 on the Pt@β-CD/C_3N_4-Ti_3C_2 was depicted (Zhu et al. 2021). Due to the black nature of Ti_3C_2, the light capture ability of C_3N_4 is significantly improved. Under the visible light irradiation, the photogenerated electrons are transferred to Ti_3C_2 quickly through layered C_3N_4, and finally transferred to Pt@β-CD NCs for reaction. In addition, due to the special hydrophobic cavity of β-CD, the Pt can adsorbed O_2 easily, which promotes the O_2 reduction and the production of H_2O_2 through two-step electron reduction. Notably, a small amount of dissolved O_2 can be directly reduced to H_2O_2 by photogenerated electrons through a two-electron reduction pathway. Moreover, the remaining H^+ can directly oxidize H_2O to O_2, thus further promote the charge carrier separation.

In summary, for the development of more efficient MXene-based photocatalysts for H_2O_2 production, the adsorption process is essential before O_2 is reduced by the photogenerated electrons, and the large two-dimensional interface and surface groups of MXene should be considered to promote the effective adsorption of the reactant. In addition, the synergistic effect of MXene and other co-catalysts can also play an important role on the photocatalytic H_2O_2 production, which may have positive significance for enhancing light-absorption capacity, optimizing carrier migration path, promoting the adsorption of reactants and desorption of products, and accelerating the rapid transformation of intermediate reactants. MXene-based photocatalysts exhibited great application prospect in photocatalytic production of H_2O_2.

9.4 PHOTOCATALYTIC STERILIZATION

Photocatalytic sterilization technology provides a new way for reducing the use of biological drugs that may have side effects on the human body. Photocatalytic sterilization mainly kills bacteria through the oxidation reaction by oxygen radicals produced under light irradiation. However, photocatalytic sterilization is faced with the problem of rapid recombination of photogenerated charge carriers. Moreover, the thermal effect generated during the irradiation process may inhibit the forward reaction. MXene, with the local surface-plasmonic resonance effect, can improve the light-absorption capacity and facilitate the separation of photogenerated charge carriers by capturing photogenerated electrons. In addition, MXene exhibit good biological antibacterial property, thus MXene can be used as an efficient co-catalyst in photocatalytic sterilization.

Li et al. prepared the Bi_2S_3/$Ti_3C_2T_x$ composite by in situ growth method (Li, Li, et al. 2021). The formation of the Bi_2S_3/$Ti_3C_2T_x$ heterostructure greatly accelerated the separation of photogenerated carriers and thus improved the production rate of reactive oxygen species (ROS). Interestingly, the ROS of Bi_2S_3 with 5% $Ti_3C_2T_x$ was significantly improved compared with that of pure $Ti_3C_2T_x$ and Bi_2S_3. In this system, $Ti_3C_2T_x$ MXene not only promoted the transfer of photogenerated carriers but also exhibited the photothermal property. During the reaction, the photogenerated electrons generated by Bi_2S_3 will transfer from the valence band to the conduction band and quickly migrate to the surface of $Ti_3C_2T_x$. In addition, the Schottky barrier formed at the interface of Bi_2S_3 and $Ti_3C_2T_x$ hinders backflow of the electrons, which enhanced the separation of the photogenerated electron-hole pairs and increased the concentration of charge carriers.

The DFT calculation results demonstrated the charge transfer mechanism between the Bi_2S_3 and $Ti_3C_2T_x$. As can be seen from Figure 9.8, for the $Bi_2S_3/Ti_3C_2F_2$ or $Bi_2S_3/Ti_3C_2O_2$, the electrostatic potential of Bi_3S_2 is higher than that of $Ti_3C_2F_2$ or $Ti_3C_2O_2$, thus forming the interface electric field and Schottky barrier at the interface, which was confirmed by the charge density difference between $Bi_2S_3/Ti_3C_2F_2$ and $Bi_2S_3/Ti_3C_2O_2$. When the interface is formed, electrons are mainly distributed at the bottom of $Ti_3C_2F_2$ or $Ti_3C_2O_2$, while the top of Bi_2S_3 becomes the electron-deficient region. In the photocatalytic sterilization, the $Bi_2S_3/Ti_3C_2T_x$ heterostructures produce more effective photogenerated electrons and holes compared with pure Bi_2S_3, which exhibited a bactericidal effect, while the photothermal effect also reduces the survival ability of bacteria. Therefore, the synergistic effect of reactive oxygen species and photothermal effects account for the high performance for the photocatalytic sterilization.

Nowadays, most MXene-based photocatalysts are mainly used in powder for photocatalytic sterilization. However, considering the practical application in sterilization, it is an effective way to design MXenes as a reactor with specific structure to promote the photocatalytic sterilization reaction. For example, Lu et al. (Lu et al. 2021) reported that a continuous-flow reactor was established to inactivate airborne bacteria by coating MXene-based photocatalysts on polyurethane foam. This reactor is suitable for photocatalytic inactivation of various infectious microorganisms under ultraviolet irradiation because $Ti_3C_2T_x/TiO_2$ can effectively destroy and inactivate airborne microorganisms. When the content of MXene was 3.4 wt%, the recombination of photoinduced electrons and holes on TiO_2 was reduced, thus increasing the activity of destroying E. coli by 30% under UV irradiation. Notably, in the absence of photocatalysis, bacteria can also be inactivated by damaging DNA due to pure UV irradiation, especially the pyrimidine dimer in DNA, but these DNA can be repaired in some cases. However, $Ti_3C_2T_x/TiO_2$ can cause physical damage to the outer membrane of microorganisms through photocatalysis, and the membrane cannot be rebuilt in the absence of nutrients, thus the bacteria will be completely killed (Figure 9.9).

MXene-based photocatalysts for photocatalytic sterilization is in its infancy stage, and more effort needs to be put into this field. For the fast development of the photocatalytic sterilization technology, the light-absorption range, efficient carrier separation, stable photothermal effect, and reactor with unique design structure should be considered for developing the MXene-based photocatalysts. In addition, it is necessary to study the mechanism of the photocatalytic sterilization. For example, the directional design of efficient MXene-based photocatalysts can be guided by identifying the free radicals that play a major role in the photocatalytic-sterilization reaction. Moreover, the effect of two-dimensional interface, surface groups, and good bacterial resistance of MXene on the performance of the photocatalytic sterilization needs to be revealed. In summary, the MXene-based photocatalysts shows a wide range of application prospects in photocatalytic sterilization, which illuminates the way for the replacement of traditional sterilization approaches in the future.

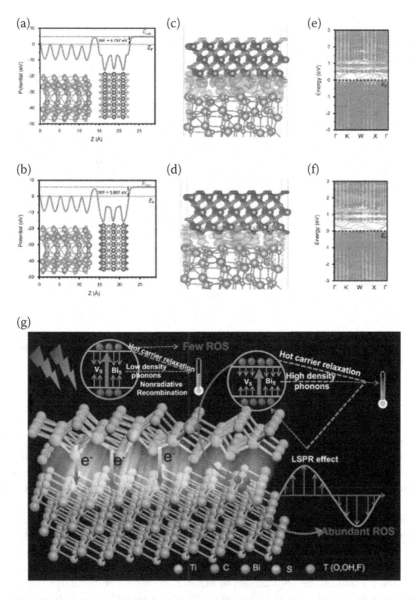

FIGURE 9.8 DFT calculation of the $Bi_2S_3/Ti_3C_2F_2$ and $Bi_2S_3/Ti_3C_2O_2$ structure. Electrostatic potential along z axis of $Bi_2S_3/Ti_3C_2F_2$ (a) and $Bi_2S_3/Ti_3C_2O_2$ (b) interface. The red dashed lines denote vacuum level (E_{vac}), the gray lines represent Fermi level (EF). The differential charge density of $Bi_2S_3/Ti_3C_2F_2$ (c) and $Bi_2S_3/Ti_3C_2O_2$ (d) interface. Pink and yellow colors denote electron excess and deficiency area, respectively. Electronic band gap structure of $Bi_2S_3/Ti_3C_2F_2$ and $Bi_2S_3/Ti_3C_2O_2$. Blue-gray colors denote contributions from the bulk part of the $Ti_3C_2F_2$ (e) and $Ti_3C_2O_2$. (f) The darkness of this color is determined by the degree of contribution. (g) Schematic diagram for photodynamic and photothermal mechanism between $Ti_3C_2T_x$ and Bi_2S_3 (Li, Li, et al. 2021).

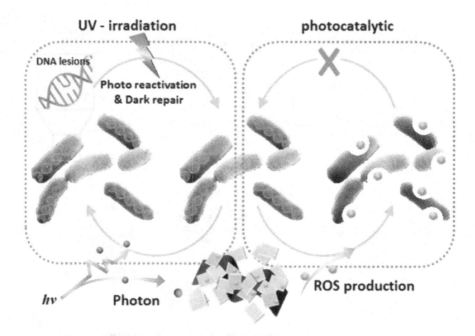

FIGURE 9.9 Diagram of inactivation and reactivation mechanism under UV irradiation and UV photocatalysis (Lu et al. 2021).

9.5 CONCLUSION AND PERSPECTIVES

In this chapter, the applications of MXenes in photocatalytic nitrogen fixation, H_2O_2 production, and sterilization are summarized. These researches on MXene greatly expanded the application range of MXenes in the field of photocatalysis (VahidMohammadi, Rosen, and Gogotsi 2021). MXenes have been widely studied in photocatalysis due to their unique character, such as the electronic conductivity, large specific surface area, two-dimensional structure, rich surface groups, and stability. In the MXene-based photocatalyst, MXenes can be used as efficient co-catalyst for the separation of photogenerated charge carriers, which exhibit the potential for replacing the precious metals in the future. However, the large-scale and low-cost production remains a major problem for the MXene-based photo-catalysts. To sum up, the applications of MXenes in various photocatalytic reactions will attract more and more attention, and we believe that MXenes will continue to make new breakthroughs in the field of photocatalysis.

ACKNOWLEDGMENTS

This work was supported by the National Natural Science Foundation of China (22078074, 21938001), Guangxi Natural Science Foundation (2019GXNSFAA-245006, 2020GXNSFDA297007, 2016GXNSFFA380015), Special funding for 'Guangxi Bagui Scholars', and Scientific Research Foundation for High-level Personnel from Guangxi University.

REFERENCES

Chen, Yiming, Wenquan Gu, Li Tan, Zhimin Ao, Taicheng An, and Shaobin Wang. 2021. Photocatalytic H_2O_2 production using Ti_3C_2 MXene as a non-noble metal cocatalyst. *Applied Catalysis A: General* 618:118127.

Elishav, Oren, Bar Mosevitzky Lis, Elisa M. Miller, et al. 2020. Progress and prospective of nitrogen-based alternative fuels. *Chemical Reviews* 120 (12):5352–5436.

Feng, Xiaofang, Zongxue Yu, Yuxi Sun, et al. 2021. Review MXenes as a new type of nanomaterial for environmental applications in the photocatalytic degradation of water pollutants. *Ceramics International* 47 (6):7321–7343.

Firestein, Konstantin L., Joel E. von Treifeldt, Dmitry G. Kvashnin, et al. 2020. Young's modulus and tensile strength of Ti_3C_2 MXene nanosheets as revealed by in situ tem probing, afm nanomechanical mapping, and theoretical calculations. *Nano Letters* 20 (8):5900–5908.

Gao, Wanguo, Xiaoman Li, Shijian Luo, et al. 2021. In situ modification of cobalt on MXene/TiO_2 as composite photocatalyst for efficient nitrogen fixation. *Journal of Colloid and Interface Science* 585:20–29.

Han, Donglin, Yajing Han, Jun Li, et al. 2020. Enhanced photocatalytic activity and photothermal effects of cu-doped metal-organic frameworks for rapid treatment of bacteria-infected wounds. *Applied Catalysis B: Environmental* 261:118248.

Hasan, Md Mehdi, Md Milon Hossain, and Hussain Kawsar Chowdhury. 2021. Two-dimensional MXene-based flexible nanostructures for functional nanodevices: A review. *Journal of Materials Chemistry A* 9 (6):3231–3269.

Kärkäs, Markus D. 2017. Photochemical generation of nitrogen-centered amidyl, hydrazonyl, and imidyl radicals: Methodology developments and catalytic applications. *ACS Catalysis* 7 (8):4999–5022.

Kim, Hyunho and Husam N. Alshareef. 2020. MXetronics: MXene-enabled electronic and photonic devices. *ACS Materials Letters* 2 (1):55–70.

Li, Jianfang, Zhaoyang Li, Xiangmei Liu, et al. 2021. Interfacial engineering of Bi_2S_3/$Ti_3C_2T_x$ MXene based on work function for rapid photo-excited bacteria-killing. *Nature Communications* 12 (1):1224.

Li, Li, Yuan Yu, Songmin Lin, et al. 2021. Single ruthenium atom supported on g-C_3N_4 as an efficient photocatalyst for nitrogen fixation in ultra-pure water. *Catalysis Communications* 153:106294.

Li, Rengui. 2018. Photocatalytic nitrogen fixation: An attractive approach for artificial photocatalysis. *Chinese Journal of Catalysis* 39 (7):1180–1188.

Li, Xibao, Weiwei Wang, Fan Dong, et al. 2021. Recent advances in noncontact external-field-assisted photocatalysis: From fundamentals to applications. *ACS Catalysis* 11 (8):4739–4769.

Liao, Yuan, Jing Qian, Gang Xie, et al. 2020. 2D-layered Ti_3C_2 MXenes for promoted synthesis of NH_3 on P25 photocatalysts. *Applied Catalysis B: Environmental* 273: 119054.

Lim, Kang Rui Garrick, Albertus D. Handoko, Srinivasa Kartik Nemani, et al. 2020. Rational design of two-dimensional transition metal carbide/nitride (MXene) hybrids and nanocomposites for catalytic energy storage and conversion. *ACS Nano* 14 (9):10834–10864.

Lipatov, Alexey, Adam Goad, Michael J. Loes, et al. 2021. High electrical conductivity and breakdown current density of individual monolayer $Ti_3C_2T_x$ MXene flakes. *Matter* 4 (4):1413–1427.

Lu, Siyi, Ge Meng, Can Wang, and Hong Chen. 2021. Photocatalytic inactivation of airborne bacteria in a polyurethane foam reactor loaded with a hybrid of MXene and anatase TiO_2 exposing {001} facets. *Chemical Engineering Journal* 404:126526.

Medford, Andrew J. and Marta C. Hatzell. 2017. Photon-driven nitrogen fixation: Current progress, thermodynamic considerations, and future outlook. *ACS Catalysis* 7 (4): 2624–2643.

Moon, Gun-hee, Mamoru Fujitsuka, Sooyeon Kim, Tetsuro Majima, Xinchen Wang, and Wonyong Choi. 2017. Eco-friendly photochemical production of H_2O_2 through O_2 reduction over carbon nitride frameworks incorporated with multiple heteroelements. *ACS Catalysis* 7 (4):2886–2895.

Peng, Jiahe, Xingzhu Chen, Wee-Jun Ong, Xiujian Zhao, and Neng Li. 2019. Surface and heterointerface engineering of 2D mxenes and their nanocomposites: Insights into electro- and photocatalysis. *Chem* 5 (1):18–50.

Qin, Jiangzhou, Xia Hu, Xinyong Li, Zhifan Yin, Baojun Liu, and Kwok-ho Lam. 2019. 0D/2D AgInS$_2$/MXene Z-scheme heterojunction nanosheets for improved ammonia photosynthesis of N$_2$. *Nano Energy* 61:27-35.

Ran, Jingrun, Guoping Gao, Fa-Tang Li, Tian-Yi Ma, Aijun Du, and Shi-Zhang Qiao. 2017. Ti$_3$C$_2$ MXene co-catalyst on metal sulfide photo-absorbers for enhanced visible-light photocatalytic hydrogen production. *Nature Communications* 8 (1):13907.

Schwaber, Mitchell J., Tali De-Medina, and Yehuda Carmeli. 2004. Epidemiological interpretation of antibiotic resistance studies – what are we missing? *Nature Reviews Microbiology* 2 (12):979–983.

Sun, Yanyan, Lei Han, and Peter Strasser. 2020. A comparative perspective of electrochemical and photochemical approaches for catalytic H_2O_2 production. *Chemical Society Reviews* 49 (18):6605–6631.

VahidMohammadi, Armin, Johanna Rosen, and Yury Gogotsi. 2021. The world of two-dimensional carbides and nitrides (MXenes). *Science* 372 (6547):eabf1581.

Wang, Hanmei, Ran Zhao, Haoxuan Hu, Xianwei Fan, Dajie Zhang, and Dong Wang. 2020. 0D/2D heterojunctions of Ti$_3$C$_2$ MXene QDs/SiC as an efficient and robust photocatalyst for boosting the visible photocatalytic NO pollutant removal ability. *ACS Applied Materials & Interfaces* 12 (36):40176–40185.

Wang, Hanmei, Ran Zhao, Junqi Qin, et al. 2019. MIL-100(Fe)/Ti$_3$C$_2$ MXene as a Schottky catalyst with enhanced photocatalytic oxidation for nitrogen fixation activities. *ACS Applied Materials & Interfaces* 11 (47):44249–44262.

Wang, Jingwen, Chaolin Li, Muhammad Rauf, Haijian Luo, Xue Sun, and Yiqi Jiang. 2021. Gas diffusion electrodes for H_2O_2 production and their applications for electrochemical degradation of organic pollutants in water: A review. *Science of The Total Environment* 759:143459.

Wang, Longwei, Xiao Zhang, Xin Yu, et al. 2019. An all-organic semiconductor C$_3$N$_4$/PDINH heterostructure with advanced antibacterial photocatalytic therapy activity. *Advanced Materials* 31 (33):1901965.

Wang, Ying and Thomas J. Meyer. 2019. A route to renewable energy triggered by the haber-bosch process. *Chem* 5 (3):496–497.

Xia, Pengfei, Shaowen Cao, Bicheng Zhu, et al. 2020. Designing a 0D/2D S-scheme heterojunction over polymeric carbon nitride for visible-light photocatalytic inactivation of bacteria. *Angewandte Chemie International Edition* 59 (13):5218–5225.

Xie, Xiao-Ying, Pin Xiao, Wei-Hai Fang, Ganglong Cui, and Walter Thiel. 2019. Probing photocatalytic nitrogen reduction to ammonia with water on the rutile TiO$_2$ (110) surface by first-principles calculations. *ACS Catalysis* 9 (10):9178–9187.

Xing, Pingxing, Pengfei Chen, Zhiqiang Chen, et al. 2018. Novel ternary MoS$_2$/C-ZnO composite with efficient performance in photocatalytic NH$_3$ synthesis under simulated sunlight. *ACS Sustainable Chemistry & Engineering* 6 (11):14866–14879.

Xiong, Jun, Pin Song, Jun Di, and Huaming Li. 2020. Atomic-level active sites steering in ultrathin photocatalysts to trigger high efficiency nitrogen fixation. *Chemical Engineering Journal* 402:126208.

Yang, Yang, Zhuotong Zeng, Guangming Zeng, et al. 2019. Ti_3C_2 Mxene/porous g-C_3N_4 interfacial Schottky junction for boosting spatial charge separation in photocatalytic H_2O_2 production. *Applied Catalysis B: Environmental* 258:117956.

Yu, Mengzhou, Zhiyu Wang, Junshan Liu, Fu Sun, Pengju Yang, and Jieshan Qiu. 2019. A hierarchically porous and hydrophilic 3D nickel–iron/MXene electrode for accelerating oxygen and hydrogen evolution at high current densities. *Nano Energy* 63:103880.

Zhang, Xinning, Bess B. Ward, and Daniel M. Sigman. 2020. Global nitrogen cycle: Critical enzymes, organisms, and processes for nitrogen budgets and dynamics. *Chemical Reviews* 120 (12):5308–5351.

Zhao, Yang, Meidan Que, Jin Chen, and Chunli Yang. 2020. MXenes as co-catalysts for the solar-driven photocatalytic reduction of CO_2. *Journal of Materials Chemistry C* 8 (46):16258–16281.

Zhou, Wei, Liang Xie, Jihui Gao, et al. 2021. Selective H_2O_2 electrosynthesis by O-doped and transition-metal-O-doped carbon cathodes via O_2 electroreduction: A critical review. *Chemical Engineering Journal* 410:128368.

Zhu, Haiguang, Qiang Xue, Guangyan Zhu, Yong Liu, Xinyue Dou, and Xun Yuan. 2021. Decorating Pt@cyclodextrin nanoclusters on C_3N_4/MXene for boosting the photocatalytic H_2O_2 production. *Journal of Materials Chemistry A* 9 (11):6872–6880.

Zuo, Gancheng, Yuting Wang, Wei Liang Teo, et al. 2020. Ultrathin $ZnIn_2S_4$ nanosheets anchored on $Ti_3C_2T_X$ MXene for photocatalytic H_2 evolution. *Angewandte Chemie International Edition* 59 (28):11287–11292.

Index

Printed in the United States
by Baker & Taylor Publisher Services